T0332514

ASSESSING FAULT MODEL AND TEST QUALITY

THE KLUWER INTERNATIONAL SERIES IN ENGINEERING AND COMPUTER SCIENCE

VLSI, COMPUTER ARCHITECTURE AND DIGITAL SIGNAL PROCESSING
Consulting Editor
Jonathan Allen

Latest Titles

ASSESSING FAULT MODEL AND TEST QUALITY

by

Kenneth M. Butler
Texas Instruments, Inc.

and

M. Ray Mercer
The University of Texas at Austin

Kluwer Academic Publishers
Boston/Dordrecht/London

Distributors for North America:
Kluwer Academic Publishers
101 Philip Drive
Assinippi Park
Norwell, Massachusetts 02061 USA

Distributors for all other countries:
Kluwer Academic Publishers Group
Distribution Centre
Post Office Box 322
3300 AH Dordrecht, THE NETHERLANDS

Library of Congress Cataloging-in-Publication Data

Butler, Kenneth M., 1962-
  Assessing fault model and test quality / by Kenneth M. Butler and
M. Ray Mercer.
    p. cm. -- (The Kluwer international series in engineering and
computer science ; #157. VLSI, computer architecture, and digital
processing)
      Includes bibliographical references and index.
      ISBN 0-7923-9222-1
        1. Digital integrated circuits--Testing. 2. Fault-tolerant
computing. I. Mercer, Melvin Ray. II. Title. III. Series: Kluwer
international series in engineering and computer science ; SECS
157. IV. Series: Kluwer international series in engineering and
computer science. VLSI, computer architecture, and digital signal
processing.
  TK7874.B85   1992
    621.381'5--dc20   91-32602
                                           CIP

Printed on acid-free paper.

Printed in the United States of America

To Stephanie, Billene, Rebecca, Elizabeth, and our parents.

Table of Contents

List of Figures

List of Tables

Preface

For many years, the dominant fault model in automatic test pattern generation (ATPG) for digital integrated circuits has been the stuck-at fault model. The static nature of stuck-at fault testing when compared to the extremely dynamic nature of integrated circuit (IC) technology has caused many to question whether or not stuck-at fault based testing is still viable. Attempts at answering this question have not been wholly satisfying due to a lack of true quantification, statistical significance, and/or high computational expense. In this monograph we introduce a methodology to address the question in a manner which circumvents the drawbacks of previous approaches.

The method is based on symbolic Boolean functional analyses using Ordered Binary Decision Diagrams (OBDDs). OBDDs have been conjectured to be an attractive representation form for Boolean functions, although cases exist for which their complexity is guaranteed to grow exponentially with input cardinality. Classes of Boolean functions which exploit the efficiencies inherent in OBDDs to a very great extent are examined in Chapter 7. Exact equations giving their OBDD sizes are derived, whereas until very recently only size bounds have been available. These size equations suggest that straightforward applications of OBDDs to design and test related problems may not prove as fruitful as was once thought.

A new functional test generation scheme utilizing OBDDs called Difference Propagation is introduced in Chapter 8. Difference Propagation enables the generation of complete test sets more efficiently than is possible with conventional test generation tools. Furthermore, it has the capacity of addressing a wider array of logical models than simply the single stuck-at fault. The application of Difference Propagation to selected fault sets and benchmark circuits has provided insights into fault model behavior. Clues are revealed as to how to best enhance fault detection using design for testability techniques.

A novel probabilistic model expressing the essential behavior of ATPG has also been developed. The model is used in conjunction with the test set information derived previously to help predict the trends of stuck-at testing with increasing circuit complexity. The results, which are detailed in Chapter 11, seem to indicate that traditional stuck-at testing methods will, below some defect level, fail to provide acceptable IC quality.

The bulk of the new material presented in this monograph comprised part of the first author's PhD research at the University of Texas at Austin. The authors gratefully acknowledge the support of this research by the Semiconductor Research Corporation under Contract #91-DP-142, by the National Science Foundation under Grant MIP-8552537, by the Innovative Science and Technology Office of the Strategic Defense Initiative Organization, administered through the Office of Naval Research under Contract N00014-86-K-0554, by the Texas Advanced Technology Program – Project No. 003658-541, and by a Departmental Grant from IBM Corporation.

The authors also wish to express their thanks and appreciation to the members of the first author's PhD Supervisory Committee, Dr. John Durbin, Dr. Joseph Rahmeh, Dr. Stephen Szygenda, and Dr. Dana Taipale. Thanks also go to the other members of Dr. Mercer's research group, Dr. Rhonda Gaede, Dr. Tom Glover, Chih-Teng Hung, Rohit Kapur, Dr. Eun Sei Park, Dr. Don Ross, and Dr. Bret Stewart. Their intellectual stimulus and generosity with software contributed greatly to the work reported here.

The first author's parents, John and Dora Butler, and parents-in-law, Albert and Shirley Watts, unceasingly lent a great deal of encouragement and support throughout his education.

Finally, and most importantly, both authors wish to acknowledge their wives, Dr. Stephanie Watts Butler and Mrs. Sharry Billene (Cannon) Mercer. Their constant provision of love, friendship, moral support, and advice made this work possible.

ASSESSING FAULT MODEL AND TEST QUALITY

Chapter 1

Introduction

The current level of complexity of integrated circuits (ICs) renders human-based test generation an extremely difficult if not impossible task. The job of test generation for mass produced ICs has thus been relegated to the domain of automated testing methods. Successful automatic test pattern generation (ATPG) requires the harmonious cooperation of two basic components: a model of physical failures called a fault model, and an algorithm capable of deriving tests for specific fault instances.

The predominant fault model used in ATPG today is the stuck-at fault model. This model, originally proposed by Eldred in 1959 [ELDR59], assumes that failures are manifested by fixed logical values at certain inputs or outputs of primitive circuit elements. At a time when digital circuits were constructed from discrete components, the stuck-at fault model made sense because actual failures often behaved in this manner. However, since 1959, VLSI circuits have come to replace discrete devices as the basic logic building blocks. Furthermore, the IC industry has proven to be extremely dynamic, evolving from bipolar to metal oxide semiconductor (MOS) devices and from multi-micron to submicron linewidths.

The imbalance of these two trends has caused IC manufacturers and researchers to question whether or not stuck-at testing is still a sensible approach to ATPG [MEI74], [WADS78a], [GALI80], [NICK80], [BEH82], [ACKE83], [LAMO83], [TIMO83], [BANE84], [BHAT84], [SHEN85], [MALA86], [FERG88], [NIGH89], [MAXW90].

Investigators of test aptness have traditionally attempted to answer one fundamental question, "Does the fault model upon which the test algorithm is based accurately mimic actual failure mechanisms?", e. g. the body of research collectively referred to as Inductive Fault Analysis [SHEN85], [FERG88].

While such work is important and useful, we propose that a second very fundamental question relating as much to the test process itself as to the model underlying it must also be answered independently of the first. This other elemental question can best be stated as, "Does the model facilitate the successful completion of the task in which it is involved?" For test generation, this is equivalent to asking if the fault model causes the test generator to select those test vectors which do the best job of locating defective circuits when they occur.

Consideration of this point amounts to evaluating the fault model as the means to an end as opposed to an end in itself. In other words, it is actually a measurement of the performance of test sets derived from the model instead of a study of the isomorphisms between physical defects and individual instances of the fault model. This objective can be achieved by studying the circumstances in which test sets generated for a *target* set of faults (often stuck-at faults) cover other *non-target* defects as well. The domain in which the target fault model performs well must certainly have ramifications to the useful life of the test methodology. Unfortunately, less work has been done to answer this question [SETH73], [CASE75], [AGAR81], [ACKE83], [HUGH86], [WOOD87], [COX88], [MILL88], [MILL89], [MIDK89], [SETH90], [PANC90], [STOR90], [SILB90], [BUTL90a], [MAXW90], [BUTL91b], [STOR91], [MAXW91]. In this research, we seek to quantify a response in a more comprehensive manner than was previously possible.

Theoretical approaches have sought to use the characteristics of the target fault sets to guarantee coverage of other types of faults [MEI74], [AGAR81]. The resulting coverage bounds have been somewhat disappointing in terms of an aggressive measure of coverage. Simulation approaches ([ABRA83], [HUGH86], [MILL88], and [MILL89]) may not obtain test sets from a large enough sample of ATPG systems to be statistically significant. Furthermore, analyzing the ensuing simulation data is a logistically challenging if not impossible task.

More recently, researchers have examined the question by collecting test statistics from actual fabricated ICs [MAXW90], [PANC90], [STOR91], [MAXW91]. Because large numbers of circuits can be examined, the results may be deemed more reliable. In the work of Pancholy et. al., the exhaustive testing of these approaches requires small circuit sizes. Furthermore, the desire for diagnosability places further constraints on the environment of the designs. The other efforts suffer from the same lack of breadth mentioned previously.

This monograph concerns a new method to ascertain ATPG test quality. In order to provide a good understanding of the motivation behind this research and the concepts upon which it is based, background is given on fault modeling, Ordered Binary Decision Diagrams, conventional ATPG techniques, integrated circuit defect levels, and test performance evaluation (Chapters 2 - 6, respectively).

1.1 Functional Test Generation Techniques

The method mentioned above is based upon complete test set generation for faults using a novel Boolean functional test generation scheme called Difference Propagation. Difference Propagation is an extension of previous research into functional test generation techniques [GAED88] which enables test set generation for a wider array of fault models. The basic concepts of Difference Propagation are treated in Chapter 8. The application of Difference Propagation to selected sets of benchmark circuits [BRGL85] and fault sets has enabled a detailed study of the functional characteristics of classical as well as emerging fault models. This data is reported in [BUTL90b] and will be discussed in Chapter 9.

1.2 Representing Symmetric Functions with OBDDs

The implementation of Difference Propagation has utilized the concept of Ordered Binary Decision Diagrams (OBDDs). OBDDs, as formalized by Bryant, are a canonical and often efficient means of representing and manipulating Boolean functions in a machine environment [BRYA86]. The basic properties of OBDDs are described in Chapter 3. Although OBDDs have long been conjectured to be a compact method for symbolic computation, there exist cases for which both the time and space complexities of the representation are guaranteed to grow exponentially with input cardinality [BRYA91], [BURC91b]. Few mathematical formulas to give the exact sizes of OBDDs for particular Boolean functions have been derived. Chapter 7 details the development of the first such size equations for classes of symmetric functions [SHAN38], [MCCL56]. Because symmetric functions exploit the more powerful of the OBDD size limitation properties to a very great extent and since the sizes are shown to be near the quadratic upper bound for many cases, it

is established that the nominal performance of the representation is likely to
be unacceptable for moderate to large sized functions [ROSS91a].

1.3   Controllability, Observability, and Detectability

Test generation consists of two basic components. To be detected, a logical fault must be excited by controlling it so that the correct binary value in
the good machine is opposite to that of the faulty machine. Furthermore, the
fault site must be made observable by at least one circuit output. For many
years research has sought to identify the roles that controllability and observability each play in rendering faults testable. Because these problems are in
general NP-complete [FUJI90], most existing data is approximate even for
small circuits [GOLD79], [SAVI84], [JAIN84], [JAIN86]. Chapter 10 concerns
the further application of Difference Propagation to this problem so that exact
measures of controllability, observability, and detectability are produced on a
per fault basis. These exact measures are then correlated to reveal interesting
relationships.

1.4   Modeling ATPG and Measuring Test Quality

As mentioned above, Difference Propagation is an integral part of a
scheme for test performance evaluation. Background on various methods
for assessing test performance is provided in Chapter 6. In an attempt to
surmount the inadequacies of the previous approaches, we have measured
"realistic" non-target defect coverage. This is accomplished by studying the
target fault and non-target defect interactions in the context of a simple but
powerful probabilistic model of the essential behavior of ATPG [BUTL90a],
[BUTL91b]. The nature of the calculations are such that they are more computationally expensive than fault simulation, but they encompass a much
wider array of test sets, thus preserving statistical significance. The ATPG
model, its assumptions, and its use are discussed in Chapter 11.

One way to illustrate this concept of an non-target defect coverage is
through the use of an "occurrence probability histogram". Assume that we
have a circuit which has N distinct possible defects, including K stuck-at
faults and $N - K$ non-stuck-at defects. In reality, N clearly tends to infinity,
but assume that in our hypothetical circuit N is a large but finite value.
Assume further that we know the probability of occurrence for each of the N
defects.

We will be using the concept of *detection probability*. Loosely stated, the detection probability of a defect is the probability that a randomly selected input vector detects the defect's presence in a circuit. Suppose that we map the integers $i, i \in \{1, \ldots, N\}$ to the N defects in order of increasing detection probability, first for the K stuck-at faults and then again for the remaining defects. That is, for either of the two sets of faults, $i < j$ iff the detection probability of defect $d_i \geq$ the detection probability of defect d_j. Figure 1.1 shows a hypothetical occurrence probability histogram for our example circuit. Because defect occurrence probabilities are presumably independent of detection probabilities, the profile of the histogram appears random.

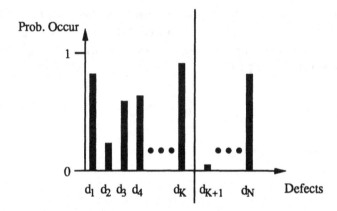

Figure 1.1: An occurrence probability histogram for a hypothetical example circuit.

Now assume that we will generate a test for each of the K stuck-at faults which appear to the left of the vertical line through the histogram. We will test only the stuck-at faults because the size of N makes it impractical to test for all defects. Furthermore, many of the non-stuck-at defects are too complex to be handled by our ATPG software. However, because some of the non-stuck-at defects have high detection probabilities and because many of them have tests in common with the stuck-at faults, we will detect some of the non-stuck-at faults "by chance". So, some of the defects to the right of the vertical line will be detected by tests generated only for defects on the left.

Depending on number of defects for which we generate tests and the magnitudes of the probabilities of occurrence of the non-stuck-at defects, we may have defects that occur which escape detection by the test vectors that we generate for the stuck-at faults. In terms of the occurrence probability histogram, the problems then become:

1. How do we measure the expected "escape rate"? Or, stated differently, how likely is it that we will "accidentally" cover all N of the defects with a test set explicitly derived for only K of them?

2. How do we reorder the indices of the defects (or in general modify our target set of faults) such that we can maximize the likelihood that all of the defects are covered by tests generated only for the target faults?

This monograph discusses efforts to answer the first question. The indications thus far have been that the traditional stuck-at testing approach will have a limited lifetime. Further research to answer the second question is in progress [KAPU90], [KAPU91].

Chapter 2

Fault Modeling

Fault models for electronic circuits are abstractions of mechanisms which could cause the device of interest to fail. As we have seen, the quality of a test set is strongly correlated with the "quality" of the fault model used. In this chapter, we review the fundamentals of fault modeling for digital circuits and discuss the characteristics of the models used in this research.

2.1 Fault Model Assumptions

When attempting to model most physical processes, some assumptions must be made to automate the solution of the problem and/or make the problem tractable. Due to the often complex nature of digital circuit failures, several assumptions are common among many fault models. The most frequently encountered of those assumptions are discussed below.

2.1.1 The Single Fault Assumption

Many test generation methods make the assumption that only a single fault is present in the circuit at any time. The main reason for the single fault assumption is due to the complexity of considering the multiple fault case. For example, a circuit with n lines can potentially contain $3^n - 2n - 1$ distinct multiple stuck-at faults. Fortunately, theoretical [AGAR81] and empirical [HUGH86] evidence has indicated that single fault test sets cover a large proportion of the multiple faults possible in a circuit, at least for the stuck-at fault case.

In the application of *on-line testing* [MCCL88], the single fault assumption implicitly contains the notion of a *frequent testing strategy*. This means that the circuit is assumed to be tested frequently enough so that any single failure is detected before a second can occur. The frequent testing assumption obviously does not hold when a single failure manifests itself as a multiple fault [ABRA80]. And, as feature sizes decrease, the likelihood of this event grows [HUGH86]. Furthermore, as the number of devices on a single chip increases, so will the probability of multiple independent failures [HUGH86]. However, if the fault coverage is in the interval [.99, 1] and the defect level is not greater than 1200 (see Chapter 5), the single fault assumption is more applicable.

2.1.2   The Time Invariance Assumption

Fluctuating power supplies and alpha particle bombardment are two examples of a host of phenomena which can cause failures whose effects are only temporary. That is, the same input vector could be applied to the circuit multiple times, and the output could be correct for some trials and incorrect for others. Failures exhibiting this time variant characteristic are called *temporary* failures and are to be distinguished from time invariant or *permanent* failures.

Due to their overwhelming predominance in manufacture testing and use by actual ATPG systems, this research has addressed solely the case of permanent failures and thus has incorporated time invariant fault models. Testing approaches have been proposed for intermittent faults and usually consist of tests for the corresponding permanent faults repeated enough times to have a high probability of detecting the intermittent faults [KAMA74], [CÔRT86].

2.1.3   The Logical Perturbation Assumption

A fault is considered *detectable* if it in some way alters the response of the machine when it is present. Certainly there are many possible detectable alterations of the normal function of an IC. These perturbations can be roughly divided into two classes, logical and parametric. A *logical* perturbation is a manifestation of a fault in which the logic value or logic function realized at some circuit node is changed. Nearly all fault models currently in use are logical fault models, and thus our work will concentrate on the performance of this type of fault model. *Parametric* perturbations are alterations in the AC characteristics of a device, such as increased current sinking or intermediate (non-binary) voltage levels. Methods exist to detect such failures which usually advocate some form of current monitoring [ACKE83], [SODE89], [MIDK89].

2.1.4 The Combinational Impact Assumption

One further simplification frequently made in fault modeling, especially for combinational circuits, is that the fault has a *combinational impact* on the circuit. By this we mean that the fault does not cause an otherwise memoryless section of circuitry to have the capacity of storing logic values. Stuck-open faults (Section 2.2.2) and bridging faults (Section 2.2.3) both have the potential of inducing memory-like behavior. The fact that stuck-at faults in general cannot induce memory behavior is another reason for their continued predominance.

2.2 Fault Model Classes

Fault models can be roughly classified according to the level of the circuit representation to which they are applied. Examples from the major classes of faults models are discussed below in increasing levels of abstraction.

2.2.1 Circuit Level Fault Models

Recent research has examined circuit level defects. These approaches draw upon the mathematical and probabilistic characteristics of integrated circuit defect sizes and distributions to systematically impose patterning defects at the mask level [STAP83], [STAP84], [FERR85]. The defects are then extracted through a series of software "filters" which eventually map the electrical impact (if any) of the faults to a level which can be analyzed further, or for which ATPG can occur [SHEN85], [FERG88], [JACO89]. This work has been very useful in deciding which types of existing ATPG fault models should be applied to various IC technologies.

2.2.2 Switch Level Fault Models

Switch level fault models assume faulty behavior which can be modeled at a transistor level circuit representation. The dominant fault models in this class are the stuck-open and stuck-closed models proposed by Wadsack [WADS78a]. These models usually assume that individual transistors are permanently fixed in either a nonconductive (stuck-open) or conductive (stuck-closed) state. They are somewhat troublesome in that stuck-closed faults may cause non-Boolean behavior and stuck-open faults often induce dynamic memory which requires two-pattern tests to detect.

2.2.3   Gate Level Fault Models

The predominant class of fault models has traditionally been the gate level fault models. Early on these models assumed purely combinational circuitry for which tests would be derived. As circuits became more complex and sequential elements were added, the combinational assumption became more questionable. *Scan* design methods, the most notable of which is Level Sensitive Scan Design (LSSD), [EICH77]), motivate continued gate level testing research. Scan designs isolate the combinational portions of circuitry by chaining the memory elements together into a *scan path*.

The Stuck-at Fault Model   Gate level fault models include the stuck-at fault model introduced in Chapter 1. The stuck-at fault model has remained popular because it maps well into many different computer representations of its binary behavior. To date, the stuck-at fault model has also been very successful at covering other types of faults as well [AGAR81], [HUGH86], [COX88], [MILL88], [MILL89]. The question of how long it will continue to perform well is one of the central points of investigation in this research.

Because so much research has been performed under the stuck-at fault assumption, many interesting properties particular to this fault model have been discovered. Some of the more relevant ones are discussed here briefly.

McCluskey and Clegg noted that certain single stuck-at faults could be divided into equivalence classes [MCCL71]. Any set of faults belonging to the same class are termed *equivalent* faults. Consider the simple AND gate example which is shown in Figure 2.1, and denote faults A stuck-at-0, B stuck-at-0, and C stuck-at-0 as A_0, B_0, and C_0, respectively. In the Karnaugh map of Figure 2.1, each input condition which is a test for one of the faults is marked with its fault designation.

The Karnaugh map indicates that the faults A_0, B_0, and C_0 all cause identical perturbations or changes in the normal function $f(C)$ realized at the output of the gate, C. Thus, faults A_0, B_0, and C_0 are said to be equivalent faults. Similar classes of faults can be found for the other primitive gates. Faults which are not at the same gate may also be equivalent, but these types of equivalencies are in general more difficult to locate. In fact, McCluskey and Clegg [MCCL71] defined three types of fault equivalence, the most general of which is applicable to all logic circuits (both combinational and sequential).

An idea similar to fault equivalence but limited to combinational circuits is that of fault *dominance*. Again, refer to Figure 2.1. Let T_{A_1} denote the set

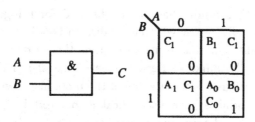

Figure 2.1: An illustration of fault equivalence and fault dominance.

of tests for the fault A_1. Let T_{C_1} be defined similarly. Since $T_{A_1} \subset T_{C_1}$, any test which detects the fault A_1 will also detect C_1. Thus, we say that fault C_1 dominates fault A_1 [SCHE72].

The ideas of fault equivalence and fault dominance can be used to reduce the number of faults in a fault set for which tests must be generated. This process of reducing the fault set is referred to as *fault collapsing*. Studies of fault collapsing techniques led to the formulation of a general set of faults, the detection of which guarantees the detection of any single stuck-at fault in a combinational circuit [BOSS71]. This fault set is called the *checkpoint* fault set. One small oversight in the theory of checkpoint fault sets was later noted and corrected [ABRA86]. The checkpoint fault set can be further reduced using the *prime* fault concept, and combinations of prime faults can represent any single or multiple fault in a circuit [CHA74]. The notion of checkpoint fault sets greatly reduces the test generation effort necessary to guarantee high coverage of single stuck-at faults and eliminates the need for explicit fault collapsing.

The Bridging Fault Model Although they appear at various levels of representation, bridging faults (BFs) are typically classified as a gate level fault model. In this model, two or more lines are postulated to be inadvertently short-circuited together. The model of the behavior of such a bridge is usually technology dependent or the result of switch-level simulation [MILL88]. The most prevalent method used is to treat the bridge as "wired logic" in the form of either a wired-AND or a wired-OR. Bridging faults exist in two varieties. Feedback bridging faults (FBFs) occur when one line is bridged to at least one other line which it feeds topologically. Such faults can induce memory, oscillatory, or non-Boolean behavior. Bridging faults that are not FBFs are called non-feedback bridging faults (NFBFs).

Many interesting results have been observed for bridging faults. Researchers have usually restricted their studies to two-line bridging faults to maintain tractability and due to the assumption that bridges of three or more lines are statistically unlikely. Recent research has shown that at least for MOS processes, bridging faults are one of the more likely types of defects to occur [SHEN85]. Studies have also indicated that certain types of bridging faults can be separated into equivalence classes and that stuck-at test sets cover a large number of bridging faults [MEI74], [ABRA83], [MILL88]. Furthermore, FBFs are typically much easier to detect than NFBFs, and ad hoc techniques such as randomly ordering the stuck-at test vectors to increase bit inversions greatly enhance bridging fault coverage [MILL88].

The Delay Fault Model Another gate level fault that is attracting increasing interest is the delay fault. Delay faults assume that the defect slows the propagation of signals through the circuit. Delay faults are detectable if the delay prevents a path from propagating signals through the circuit at the system clock rate [LESS80]. They are often modeled as either the *gate delay model* or the *path delay model*. The gate delay model assumes a lumped delay at gate inputs and outputs. The path delay model does not assume a lumped delay, but instead circuit paths are tested at the operational system clock interval [SMIT85]. More recently, a hybrid model combining elements of the gate and path delay models has been proposed [PARK87]. Much like stuck-open faults, delay faults require multi-pattern tests for detection. This presents a problem in that conventional latch designs in LSSD environments do not allow arbitrary two-pattern tests to be applied. At least one low overhead double latch design has been proposed which facilitates the application of deterministic two-pattern tests [GLOV88].

2.2.4   Other Fault Models

Higher levels of fault model classes include Boolean level models [FRID74], functional level fault models [SILB87], [THAT80], [BRAH84], register transfer level fault models, and behavioral level fault models. The vast number of working fault models prohibits an extensive review. Fault models in these classes are often similar to the lower level models discussed above, except that they must be increasingly abstract to accommodate the increasingly abstract primitives in these representations.

Chapter 3

Ordered Binary Decision Diagrams

In order to facilitate a detailed study of the perturbations of various fault models on the normal functioning of a given circuit, it is helpful to have the capacity to find all the tests for each fault in a fault set. One procedure to gather this information would be to inject each fault in the fault set, one at a time, and simulate all possible input patterns, noting when departures from the good machine outputs occur for each fault. An exhaustive method similar to the one just described was proposed in [BEH82]. Obviously, the time required for exhaustive approaches can become prohibitive quickly as circuit sizes grow.

A more efficient means for generating the same information is through the use of Ordered Binary Decision Diagrams. The concept of OBDDs and their various applications are described below.

Portions of this chapter and much of Chapter 7 are taken from the paper "Exact Ordered Binary Decision Diagram Size When Representing Classes of Symmetric Functions", by Don E. Ross and the authors, which will soon appear in the *Journal of Electronic Testing: Theory and Applications (JETTA)*, Volume 2 No. 3, Kluwer Academic Publishers.

3.1   History of OBDDs

Binary decision diagrams were first introduced as a representation for switching functions by Lee [LEE59]. They were further refined by McCarthy [MCCA63] and independently by Akers [AKER78b]. Formalization and improved algorithms were provided by Bryant [BRYA86]. Bryant also showed that OBDDs are well-defined and canonical [BRYA86], [BRAC90].

The properties of OBDDs have been carefully studied and many applications for them have been identified [MCCA63], [PAYN77], [AKER77], [AKER78b], [AKER78a], [CERN79], [MORE82], [SILV85], [CERN85],

[BRYA86], [NAIR86], [SUPO86], [CHAN86], [ABAD86], [STAN88], [FUJI88], [MALI88], [MADR88], [GAED88], [BERM88], [SIMO88], [MCGE89], [CHO89], [DEVA89], [FRIE90], [BUTL90a], [BUTL90b], [BRAC90], [CLAR90], [ISHI90], [MINA90], [SATO90], [BUTL90c], [ROSS90], [CERN90], [FUJI90], [LIN90b], [SRIN90], [COUD90], [TOUA90], [OHMU90], [MATS90], [LIN90a], [BRYA91], [ROSS91b], [BUTL91b], [HAMI91], [KAPU91], [BUTL91a], [BURC91a], [BURC91b], [CHEN91], [DEGU91], [DEVA91], [ERCO91], [JAIN91], [JU91], [NAJM91], [OCHI91]. A function is pictorially described using a directed acyclic graph which has certain properties. A vertex in an OBDD corresponds to the occurrence of a literal in an implicant or implicate of a function, and a labeled edge exiting a vertex represents the binary value to which the literal corresponding to the vertex has been set. An example of an OBDD for the function $f = AB + CD$ is shown in Figure 3.1.

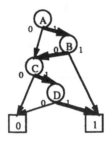

Figure 3.1: An OBDD for the function $f = AB + CD$.

OBDDs are conventionally drawn vertically with the edges directed downward. The uppermost vertex of a graph drawn in this fashion is called the *root*. The two lowest vertices are called *terminal vertices* or *termini* and have a special meaning, to be described below. A path through an OBDD from the root to a terminal vertex corresponds to a set of input conditions to the switching function. The binary value of the terminus at which the path ends is the value of the function for that set of input conditions. For example, for the input condition $A\overline{B}CD$, the corresponding path through the OBDD is shown with heavy lines in Figure 3.1. Because the path ends at the 1 terminus, the value of the function for this input set is 1.

3.2   Properties of OBDDs

OBDDs as defined by Akers [AKER78b] and formalized by Bryant [BRYA86] adhere to the following six properties [ROSS91a].

Property 3.1 *The OBDD represents the entire function of n variables, and is a rooted, acyclic digraph. Any vertex in the graph other than the root can be considered to root a subgraph which is itself an OBDD representing a residual function of $n - 1$ or fewer variables.*

Property 3.2 *Each vertex in the graph represents an input variable, x_i, and has two edges, one for $x_i = 0$, the other for $x_i = 1$. These edges each point to the residue created when x_i is assigned the value indicated on the edge, e. g. a vertex representing input x_i and function $f(x_i, x_{i+1}, \ldots, x_n)$ has its 0-edge pointing to an OBDD representing function $f(x_i = 0, x_{i+1}, \ldots, x_n)$, while its 1-edge points to an OBDD representing function $f(x_i = 1, x_{i+1}, \ldots, x_n)$. Note that these residues are functions of $(x_{i+1}, \ldots, x_n)$, one less variable than $(x_i, \ldots, x_n)$.*

Property 3.3 *The function is expanded about each of its variables in a pre-specified order. If the ordered inputs are relabeled $(z_1, \ldots, z_n)$, then there are n! possible unique relabelings, one for each of the one-to-one and onto mappings from the original input set $\{x_1, \ldots, x_n\}$ to the relabeled input set $\{z_1, \ldots, z_n\}$. The relabeled index, i for z_i, will determine the expansion order of the original input variable x_1 for ordering $< x_1 \Rightarrow z_i, \ldots, x_n \Rightarrow z_k >$. The input corresponding to z_1 is expanded first, then z_2 is expanded (within each of the residual functions created by the expansion about z_1), ..., until z_n is expanded as the final residue for all expansion paths. See Figure 3.2 for an example of two OBDDs representing the same function, but under different mappings (orderings). Note that these two graphs are nonisomorphic, due to the labelings on the directed edges.*

Property 3.4 *The final residues, $f = 0$ and $f = 1$, are represented as special termini of the OBDD, and can be considered as the last vertices in the ordering, with ordering index $n + 1$.*

Property 3.5 *If a residue (including the original function) is independent of input variable x_i, then no vertex representing x_i will appear within the OBDD representing that residue. In graphical terms, if both edges of an OBDD vertex point to the same successor vertex, then the vertex with the equivalent edges is deleted from the OBDD, and all incoming edges to the deleted vertex are redirected to its successor vertex.*

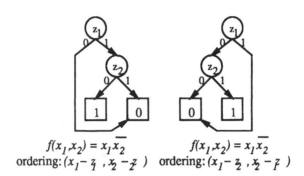

Figure 3.2: The OBDDs for $f(x_1, x_2) = x_1 \bullet \overline{x_2}$ for all 2! orderings.

Property 3.6 *Every residue which exists in an OBDD must be unique. If the functional expansion creates two residues which are equivalent, only one of them will be represented within the OBDD. In graphical terms, this means that no isomorphic subgraphs can exist in an OBDD. If a graphical expansion creates isomorphic subgraphs, all but one of the isomorphic subgraphs must be eliminated, and all edges which pointed to any of the original isomorphic subgraphs must be redirected to point to the one remaining, representative subgraph.*

OBDD sizes vary with the complexity of the functions being represented as well as variable ordering. The problem of finding an optimal ordering is known to be difficult [BRYA86], [FRIE90], but recent work has produced some useful heuristics [FUJI88], [MALI88], [MALI90], [BUTL91a]. While OBDDs are often a time and space efficient means of manipulating switching functions, their size has exhibited exponential growth for certain functions, such as integer multipliers [BRYA86], [BRYA91], [BURC91b].

3.3 Shannon's Expansion Theorem

As stated in [BUTL87], OBDDs can be viewed as a pictorial interpretation of Shannon's Expansion Theorem [SHAN38]. The x_i-residue of $f(x_1, \ldots, x_i, \ldots, x_n)$ is $f(x_1, \ldots, x_{i-1}, 1, x_{i+1}, \ldots, x_n)$. Similarly, the $\overline{x_i}$-residue is $f(x_1, \ldots, x_{i-1}, 0, x_{i+1}, \ldots, x_n)$. Shannon's expansion theorem states that any switching function $f(x_1, \ldots, x_n)$ can be represented in terms of the residues of any of its variables. Formally, this theorem is

$$f(x_1, \ldots, x_i, \ldots, x_n) =$$
$$x_i f(x_1, \ldots, x_{i-1}, 1, x_{i+1}, \ldots, x_n)$$
$$+\overline{x_i} f(x_1, \ldots, x_{i-1}, 0, x_{i+1}, \ldots, x_n). \tag{3.1}$$

OBDDs express these residue relationships in that edges from each vertex point to the residues obtained from that function by assigning the corresponding edge values to the variable associated with the vertex.

Chapter 4

Automatic Test Pattern Generation

One goal of this research has been to produce a viable measure of non-target defect coverage of test sets for target faults obtained from an ATPG system. Because many ATPG algorithms exist, each guided by different heuristics with varying behavior, it has been necessary to capture the essential components of these algorithms in a model of the process. This model will be discussed in detail in Chapter 11. In this section we concentrate on describing the common characteristics of conventional ATPG systems. The discussion will address solely techniques for test generation at the logic gate level.

4.1   ATPG Problem Specification

The ultimate task that ATPG must accomplish is to derive a set of test patterns which, when applied to a circuit under test (CUT), have a high likelihood of declaring the circuit faulty if it is indeed flawed. Certainly if all possible failure mechanisms could be detected by a given test set then the likelihood described above would be 100%. Unfortunately, even for the simplest circuits this is theoretically (and quite often practically) impossible. Defects could always exist that require increasingly longer series' of patterns for their detection. However, it is also true that the increasing complexity of defects that would require such long series' of patterns to be detected would make them less and less likely to actually occur.

Thus, we can approximate all possible defect mechanisms by all likely defect mechanisms. For most circuits, the set of all likely defect mechanisms is probably still too large to explicitly address each member. So, we select a set of candidate failure mechanisms which are "straightforward, accurate, and easy to use" [HAYE85]. This set of failure mechanisms, or faults, is the target fault set.

4.2   Conventional ATPG Algorithms

Having selected the target fault set, a test set must be derived which detects occurrences of these faults. The hope is that in doing so, the test set which also detect many other types of (non-target) defects which might also occur. Practice has shown that this is often the case, as was stated in Section 2.2.3.

The approach of random test generation, where circuit inputs are selected pseudo-randomly and applied to the CUT, is one vehicle for generating test patterns. Work in this field has resulted in the observation that very few random test patterns can achieve target fault coverages that are quite high [AGRA75], [DAVI76], [DAVI81]. Therefore, many typical faults have relatively large test sets, and usually some small subsets are more difficult to test. These hard-to-detect, or *random pattern resistant* faults preclude random test generation from being a definitive solution to the testing problem. So, determinism must often be introduced at some point. To minimize test pattern length, deterministic test pattern generation can be used for all faults in the target set.

4.2.1   Fault Excitation and Observation

To achieve successful ATPG, two criteria must be satisfied simultaneously. First of all the fault must be *excited*. A logical fault has been excited if a condition exists where the logic value at the fault site is different than it would be in a fault-free circuit. In order to be detected, the fault site must also be *observable*. That is, a path must exist between the fault site and some output point which can be measured or observed. In this light, the process of deterministic test pattern generation can be seen to be very similar to computational search procedures. Conditions are sought in which both excitation and observation are satisfied. For general sequential testing, this problem is extremely difficult because of the additional dimension of time. Scan-based designs ease the problem somewhat by removal of the time dimension, although deterministic combinational test pattern generation is still known to be an NP-complete problem [IBAR75].

4.2.2   Line Implication and Justification

From this point, the solutions of ATPG begin to diverge somewhat [ROTH66], [GOEL81], [FUJI83], [KIRK87], [SCHU88b], [SCHU88a],

[GIRA90]. Because most conventional systems are based on the stuck-at fault model, we will couch our discussion in those terms. To excite a stuck-at fault, the binary value opposite to the stuck-at fault value is imposed on the faulted line. Generally, the next step in the process is *implication*, where all values implied by the fault excitation conditions are specified, both upstream and downstream of the fault site. For example, if a stuck-at-0 fault on the output of an AND gate is the target fault, then a binary value of 1 is needed on that line to excite the fault, which implies that all inputs to the AND gate must have the value 1 in order to produce the fault excitation.

Implication is accompanied by *justification*. Values assigned to lines during the implication phase must be justified by other values in the driving circuitry. In the special case of a faulted line, such as at the output of the AND gate in our previous example, a D $(\overline{D})$ is used to represent the presence of a 1 (0) in the good circuit and a 0 (1) in the faulty circuit. Several multi-valued algebras have been proposed to perform these calculations depending on the specific testing application, the first of which appeared in [ROTH66].

The example cited above was trivial in that implication completely specified the decisions that had to be made. Often the choice is not as clear, as in the case of "implication" of a 0 on the output of an AND gate. In these cases, decisions must be made as to which driving or driven lines should have which values. This is a crucial point in that ATPG algorithms must efficiently *prune* the search space. These decisions are usually based on a variety of heuristics having to do with controllability, observability, and testability measures.

4.2.3 Backtracking

Occasionally during either implication or justification, conflicts will arise on various lines. A conflict has occurred when two contrasting values are simultaneously required on one line to satisfy two different objectives. In this case, the search process is altered by a process known as *backtracking*. In backtracking, decisions made during prior justification are reversed so that other portions of the search space can be investigated. This procedure is often fruitful except in the cases of random pattern resistant or *redundant* faults.

Redundant faults are faults for which no test exists. An example of a redundant fault is shown in Figure 4.1. The single stuck-at-0 fault on the output of the XOR gate is an uncontrollable redundant fault [RATI83]. That is, the fault excitation condition of 1 can not be justified as the quiescent state at the output of the XOR gate.

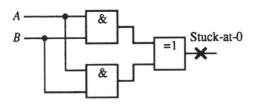

Figure 4.1: An example of a redundant fault.

By the very definition of redundancy, the search space must be completely exhausted for a fault to be proved redundant. These types of faults can lead to exorbitant amounts of computation time wasted in fruitless searches. Such problems are often circumvented by setting a *backtrack limit*. When the backtrack limit is exceeded, the fault is *aborted* and no test is specified for it. The problem of redundant faults has traditionally been a cause of great concern, although recent conventional ATPG systems have reported good results in locating redundant faults with low backtrack limits [SCHU88b], [SCHU88a], [GIRA90].

4.3   Boolean Functional Test Generation

As we saw in the last section, proving a fault redundant requires the exhaustive search of large spaces. Depending on how efficiently the conventional ATPG algorithm prunes the search space, a large amount of computation time can be expended due to the serial nature of its backtrack-based approach. One way to solve this problem would be to "deserialize" the algorithm so that large spaces can be probed concurrently. This is the central theme behind Boolean functional test generation schemes. In this section we describe the basic makeup of Boolean functional test generation.

4.3.1   Functionally Describing Controllability and Observability Information

In information-intensive applications such as redundancy proving, conventional ATPG systems treat the generation of the controllability and observability information of a fault as a binary search procedure. Because the method amounts to little more than a constrained simulation problem, these algorithms can examine individual states of the combinational machine very quickly. However, the very large size of some search spaces creates problems for a serial approach.

By examining each input condition (or small bundles of input conditions) for a solution, we are in some sense trying every minterm (implicant) of a Boolean function to see if it "works". If the input space and problem constraints could somehow be described functionally, then perhaps the constraints of the problem could be imposed on the function space such that the solution or set of solutions could be found in one operation. In the case of combinational ATPG, the input space is simply the Boolean functions realized by the circuit. The problem constraints are that a logical fault forces some perturbation somewhere in the circuitry that may (detectable fault) or may not (redundant fault) change the observable logical functionality of the circuit.

The approach in Boolean functional test generation would proceed as follows:

1. For all "necessary" lines (to be defined later), derive information which describes the normal "good" functions at those lines.

2. Using the information derived in Step 1 and for each fault of interest:

   (a) Derive further information which describes all conditions which excite the fault.

   (b) Also derive information describing the conditions which cause the effect of the fault to be observable at one or more output points that can be directly monitored.

   (c) Check the intersection of the controllability information and the observability information to see if a condition exists in which the fault is both excited and observable. If the answer is yes, the fault is detectable. If no, the fault is redundant.

For most typical fault sets, all the lines in the circuit would be necessary. An example of a case where not all lines would be necessary is shown in Figure 4.2. The dark square represents a single fault site within the circuit. The triangular shaped region to the right of the fault is the circuitry which is directly or indirectly reachable from the fault site. The hypothetical fault feeds the three uppermost circuit outputs. The dark region represents the union of the circuitry which feeds those particular outputs. In this simple case, functional calculations would be necessary only for those gates falling within the shaded region. All other circuitry could be neglected. For a set

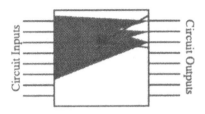

Figure 4.2: An example illustrating the functional information needed for a particular fault configuration.

of individual faults, a similar topological analysis would occur for each fault, and calculations would occur for the union of the sets of circuitry.

Representing information functionally instead of in the Boolean domain allows the concurrent generation of all possible tests for a fault, not just one or a few. In typical test generation applications, such information is more than the user may want. This research is one instance in which complete test sets are required for accurate calculations, as will be seen in later chapters.

4.3.2   CATAPULT

CATAPULT (Concurrent Automatic Testing Allowing Parallelization and Using Limited Topology) was an early effort at using an OBDD-based functional approach to generate tests for faults in combinational circuits [GAED88]. CATAPULT combines the basic components of Boolean functional test generation with the Boolean difference [AKER59] to derive partial or complete test set information.

CATAPULT begins by preprocessing the input circuit to find the set of lines for which functional information is necessary. Normally, only the circuit inputs would be represented as switching variables in the calculations. However, the use of the Boolean difference requires that certain other points in the circuit must be represented as well. These points are a subset of the circuit *fanout stems*. Fanout stems are inputs lines or gate outputs which "fanout" to feed more than one other gate input. The gate inputs to which the stem fans out are referred to as *fanout branches*.

Test generation continues by deriving the "good function" information for the lines indicated during preprocessing. Next, beginning at the outputs and working backwards, CATAPULT calculates observability information for each line along the path to fault sites. Relatively simple conditions exist by which the observability of a gate input can be calculated from the observability function of its output and the controllability functions of its other inputs. However, the same is not true of the observability of a fanout stem. It cannot be easily calculated from the observabilities of its branches. The Boolean difference provides a mechanism for calculating stem observabilities. Consider Figure 4.3. If this stem is the only one for which we will calculate the Boolean difference, then the switching variables in which the output functions are expressed are the input variables (represented as X) and z, the stem variable. Using the notation of the Boolean difference, the observability function of stem z with respect to the output $f(X, z)$ is

$$\frac{df}{dz} = f(X, 0) \oplus f(X, 1). \tag{4.1}$$

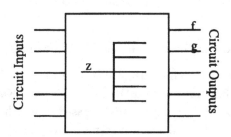

Figure 4.3: An example illustrating the use of Boolean difference to calculate stem observability functions in CATAPULT.

If the stem feeds output g, then $\frac{dg}{dz}$ must also be calculated and ORed with $\frac{df}{dz}$. The same is true for all the remaining outputs. Thus, the Boolean difference requires that functions be described in terms of not only input variables but certain fanout stem variables as well. The needed fanout stem variables are calculated along with the necessary gates during the preprocessing phase of computation.

CATAPULT continues alternately using the Boolean difference and backward gate propagation to calculate observability information at all points where it is required. The final step in the process is then to intersect the controllability and observability functions as fault sites are encountered. Depending on the fault location, this may require composition or *partial composition* [GAED88] to remove the non-input variables from the test OBDD. The result is that the test OBDD either reveals at least one test input in the case of a detectable fault or the entire OBDD collapses to a 0 terminus for a redundant fault.

Because redundant faults are particularly difficult for conventional ATPG systems, they are typically aborted during test generation runs. However, random pattern resistant faults, which are actually testable but with very small test sets, are sometimes aborted as well. Thus, it is not true that a fault aborted by a conventional test generator must be redundant. This is where a Boolean functional tool such as CATAPULT becomes useful. CATAPULT was run on specifically those faults that were aborted by TOPS [KIRK87] to decide whether or not they were redundant. By setting backtrack limits very high, TOPS was allowed to attempt to exhaustively probe the search space for the same answers. In some cases, TOPS won, but CATAPULT did post a much faster time for redundancy identification in one of the more difficult examples of redundancy in the benchmark circuit set.

4.3.3 Other Boolean Test Generation Techniques

At least two other systems exist which have the capacity of test generation using Boolean functions and OBDDs. Stannard and Kaminska [STAN88] published an abstract concerning an OBDD-based approach for detecting hard faults in combinational circuits which seems similar to CATAPULT. Unfortunately, the abstract is too brief to provide a detailed description of the theoretical background and implementation.

Cho and Bryant have developed an extension to the COSMOS [BRYA87] switch level simulator employing an approach they call *symbolic fault simulation*. Their system is significant in that it addresses many aspects of switch level behavior, such as bidirectionality, charge sharing, signal strength, etc.

In their approach to symbolic fault simulation, a Boolean description of good and faulty machines are derived using COSMOS. Efficient manipulation of OBDDs is then used to find the differences between the good and faulty machines. Finally, conditions are sought which satisfy these differences for some possible function of both inputs and state. Although the application environment is somewhat different, this approach to test generation is similar to the one employed in this research. The details of our approach will be given in Chapter 8.

Chapter 5

Defect Level

Consumers of large quantities of integrated circuits usually consider the reliability of the parts to be a major factor when selecting a manufacturer from whom they will purchase the devices they need. Specifically, if some consumer determines that the mass production of a system will require exactly 1000 copies of an IC, they obviously desire to be shipped at least 1000 functional chips. IC manufacturers generally protect themselves by shipping more parts than were ordered, knowing that since the production testing of the parts was not exhaustive, some parts may have been declared "good" when they were actually faulty. Because this issue of "quality level" is intimately tied to the level of testing to which the parts were subjected and the quality of the test itself, researchers have sought to more formally define quality level. The goals of this research are to study fault models and the quality of test sets that they provide. Thus, one possible benefit of this work is a more effective use of testing to screen out bad parts before they are shipped to the customer. It is therefore proper to review the more interesting developments in IC quality research.

5.1 Definition of Defect Level

The notion of the proportion of truly good parts among those which pass all phases of production testing is called the *quality level*, QL, of the lot [MCCL88]. Obviously, $0 \leq QL \leq 1$. Conventionally, IC manufacturers characterize their shipment quality by the complement of QL which is called the *defect level*, DL. By the complement, we mean that $DL = 1 - QL$. Defect levels are generally quoted as the average number of defective parts per million parts, or DPM.

Using quality level and defect level as defined in the preceding paragraphs, it is impossible to empirically determine the defect level for VLSI circuits. This is because for any given test set, there can always exist at least one defect which cannot be detected by the test set. Thus, several researchers have attempted various estimates of defect level [MCCL88], [WADS78b], [GRIF80], [WADS81], [WILL81], [SETH84], [PARK89], [SAVI90], [WILL90].

5.2   Defect Level Simplifying Assumptions

The typical assumptions made in defect level modeling include the following:

1. Spot defects, small variations in one or more layers of the integrated circuit, occur with an equally likely random distribution across the surface of a wafer. (Uneven distributions are modeled in [EICH83].)

2. Spot defect occurrences are independent events.

3. The yield can be approximated by the fraction of devices which pass production tests. Research into mathematical yield models includes [STAP75], [OSBU88].

4. The spot defects are small enough to manifest themselves as a single fault ([MCCL88]and [SETH84] address multiple manifestations by a single defect.)

5. Defects in circuit processing are the largest contributor to the defect level. Other factors include improper contact bonding and/or encapsulation, shipping and handling stress, static electricity, failure after functional test, etc.

5.3   Defect Level Models

An early effort was made by Wadsack to describe defect level as a function of yield [WADS78b]. Wadsack assumed that the number of faults on a chip was bounded by some integer n. He further assumed that the numbers of chips with i faults, $1 \leq i \leq n$, were geometrically distributed, and he derived the defect level equation

$$DL = (1 - y)(1 - T), \qquad (5.1)$$

where $T =$ the fault coverage of single stuck-at, stuck-open, and stuck-closed faults and Y is the yield of good devices as estimated from the fraction of devices passing production testing. His results did not match extremely well with experimental data. Wadsack hypothesized that the observed discrepancy might be due to the fact that detection of stuck-open faults is dependent not only on the test vectors, but also their ordering, and/or analog effects [WADS78b].

A slightly different approach was taken by Wadsack in a second paper [WADS81]. Here he modeled uneven defect distributions by a gamma distribution and he chose a binomial distribution to approximate chip defect counts. This time the derivation of the defect level equation resulted in

$$DL = 1 - \left(\frac{1 + \beta T}{1 + \beta}\right)^{\alpha}, \tag{5.2}$$

where α and β are factors which account for the unevenness in the defect distribution and the IC yield was found to be $Y = \left(\frac{1}{1+\beta}\right)^{\alpha}$. Empirical observations showed good agreement with the theory.

A frequently referenced correspondence on this subject is that of Williams and Brown [WILL81]. Their initial assumptions were generally the same as those of Wadsack's second work, except that they used a random equiprobable defect distribution. This resulted in the famous defect level equation

$$DL = 1 - Y^{(1-T)}. \tag{5.3}$$

where T is usually defined as the proportion of single stuck-at faults covered by the test. No experimental results were reported in this paper, and to date little empirical data has been published on this topic, although one paper appeared shortly before the Williams and Brown correspondence was published [HARR80]. It is interesting to note that under the constraint of a constant Y,

$$\lim_{\alpha \to \infty} 1 - \left(\frac{1 + \beta T}{1 + \beta}\right)^{\alpha} = 1 - Y^{(1-T)}. \tag{5.4}$$

An interesting extension to the model of [WILL81] appeared recently in [MCCL88]. Here McCluskey and Buelow noted that the assumed one-to-one correspondence between defects and faults may not be valid. They also pointed out that most of the previous work was based on lots with relatively low yield. For high yield lots, simplifications can be made in the defect level equation of [WILL81]. They defined *test transparency* (TT) as the number of physical defects that escape detection during functional test and claimed that this quantity can be approximated by $1 - T$. Also, DPM values ≤ 1000 permit further simplifications. Their work resulted in the equation

$$DL = TT(-lnY), \qquad (5.5)$$

and for $Y \geq 90\%$, they noted that $-lnY \approx 1 - Y$, so we have

$$DL = TT(1 - Y). \qquad (5.6)$$

If we choose to approximate TT by $1 - T$ as suggested, Equation (5.6) becomes Equation (5.1). Thus, McCluskey and Buelow claim that for high yield and low defect level lots, the result of Wadsack's original work [WADS78b] is a good approximation. Experiments with actual lots showed some slight disparity between theory and results which were attributed to the fact that 99.9% single stuck-at fault coverage was assumed to detect all failures considered [MCCL88].

Chapter 6

Test Performance Evaluation

In order to better understand and continually improve the manufacturing test process, it would be helpful to have a definitive measure of the "quality" of the test sets used. One method of measuring test quality might be to simply monitor the *field reject rate* - the number of parts returned or reported to the manufacturer as faulty. However, this is a crude measure for several reasons. Circuits can be damaged by a number of post-manufacture phenomena such as electrostatic discharge effects, poor encapsulation or wire bonding, etc. Also, on very complex circuits, some faults may never be detected if the corresponding portion of circuitry is not used by the consumer.

Most work in this area has focused on the fault models upon which the ATPG system is based. The motivation here is the belief that a better understanding of the characteristics of the fault models we now use will help us either tune them or use them in such a way as to realize higher quality test sets. While this may be true, ranking failure mechanisms will give no information about whether or not test sets derived from the abstract fault models will detect the failures. In other words, we will know nothing about the quality levels of the ICs emerging from the tester.

A much better way to measure test performance is to concentrate on the test generation process itself, or some aspect of it. By doing so we can discover how the test process interacts with the fault model to cause not only instances of the model to be detected, but other fault types as well. In this section, we survey the relatively small number of approaches to this problem presented in the literature.

6.1 Theoretical Approaches

Early approaches used the properties of the target (single stuck-at) fault model to guarantee coverage of certain types of non-target defects [MEI74],

[AGAR81], [ABRA83]. While these bounds help clarify the picture some-what, the demand for higher quality testing fosters a need for a more rigorous measure of test performance.

6.2 Fault Simulation Approaches

The application of fault simulation is another method by which test qual-ity can be quantified [HUGH86], [MILL88], [MILL89]. In this approach, test sets are sampled from various test generation systems. Each of the test sets are fault simulated and the coverage of non-target defects is measured. After simulation of all the test sets, aggregate statistics are compiled to give an overall coverage ratio. This method also provides the capacity of comparing different test generation algorithms to see which has better non-target defect coverage statistics.

There are some disadvantages to this approach, however. First of all, as we saw in Section 4.2, many typical target faults are very detectable. Thus, there are often a very large number of possible test sets that guarantee very high or 100% coverage. The sample size of test sets would have to be very large in such cases to ensure a proper level of statistical significance. Moreover, the computational expense of repeated fault simulation for a large number of target fault test sets and non-target defects renders application of the method to large circuit sizes and circuit sets to be prohibitive.

6.3 Test Application Approaches

In the work of Maxwell and Wunderlich [MAXW90], a small series of Programmable Logic Arrays (PLAs) are subjected to several different test sets. Then, the numbers of circuits which pass some test sets and fail others are compared to see if certain types of test sets perform better than others. In following work [MAXW91], Maxwell and Aitken examined the quality levels arising from various test procedures for a 10,000 gate standard cell device. One significant result of this research was that lower defect levels were realized with "functional" patterns than with ATPG-derived scan test vectors which had higher stuck-at fault coverage. This experimental technique can provide quite useful information. Unfortunately, it is a laborious task to carry out in a systematic and statistically significant way.

Pancholy, et al., have designed a combinational circuit and an accompanying robust test scheme which has extremely good diagnostic properties [PANC90]. Their method allows them to identify precisely which types of non-target defects escape detection by target fault test sets. In the rare instances where the results are inconclusive, they employ an electron-beam voltage-contrast machine to physically diagnose the failure. The results of their research strongly support those which will be stated in later sections. The one drawback of this method is that the need for high diagnosability constrains the design to be quite small and free of reconvergent fanout, which is atypical of most designs.

6.4 Layout Driven Approaches

Finally, the circuit level fault model studies discussed in Section 2.2.1 constitute a "back door" approach to improving test quality. If for a given technology the realistic faults can be shown to map well to specific fault models, then the performance of test generation based on those types of models can be both observed and maximized.

Chapter 7

OBDDs for Symmetric Functions

Much of the analysis reported in this monograph has been achieved through the symbolic representation of Boolean functions. OBDDs have been conjectured to be an economical vehicle for Boolean functional manipulation, but just how efficient are they? There is no known canonical Boolean function representation which remains tractably bounded for any arbitrary switching function. However, the OBDD is linear in size for some functions which cannot be represented as compactly in other canonical forms. The exclusive-OR (XOR, $\oplus$) is an example of such a function. OBDD size has also been shown to grow exponentially for some functions regardless of the variable ordering. An example of this class of circuits is the integer multipler [BRYA91], although recent research has demonstrated techniques to address the problem [BURC91b].

The size of the representation is important since it is directly proportional to the amount of computer memory required to perform calculations. However, because the time complexity of OBDD operations is also proportional to the size of the operand OBDDs [BRYA86], the representation size is paramount. Thus, in trying to quantify just how useful OBDDs are (and will be), one would best proceed by trying to calculate OBDDs sizes for functions with particularly "good" or "bad" characteristics. In this section, we discuss the details of some derivations of exact OBDD size equations.

7.1 Symmetric Functions

Symmetric functions are a special class of Boolean functions with some very useful properties. Boolean functions can be *partially* symmetric or *totally* symmetric. A Boolean function is totally symmetric iff it is unchanged by any permutation of its variables [MCCL86]. If we define a function $f(x_1, \ldots, x_n)$ of n variables, then f is partially symmetric in $x_{i1}, x_{i2}, \ldots, x_{im}$ iff it is unchanged by any permutation of variables $x_{i1}, x_{i2}, \ldots, x_{im}$. For the remainder of this chapter, we will discuss functions of n variables which are totally symmetric.

Shannon [SHAN38] developed an integer representation for symmetric functions which can represent any arbitrary n input symmetric function using at most $n + 1$ integers, and which is therefore linearly bounded in size for any arbitrary symmetric function. However, this representation can represent symmetric functions only. Shannon's representation can be best understood by considering the results of the definition of total symmetry. Because the functional value cannot be changed by permuting any variables, then it actually does not depend on *which* variables are true, but *how many*. Thus, a symmetric function can be uniquely represented by a list of at most $n + 1$ integers which specify the counts of true variables that imply the function. This list will hereafter be called the M-set of a function. Some common symmetric Boolean functions expressed as M-sets are: $f=0$, $M = \{\}$; $f=1$, $M = \{0, \ldots, n\}$; $f=\text{OR}$, $M = \{1, \ldots, n\}$; $f=\text{AND}$, $M = \{n\}$; $f=\text{NOR}$, $M = \{0\}$; $f=\text{NAND}$, $M = \{0, \ldots, n-1\}$; and $f=\text{XOR}$, $M = \{i \mid i \text{ is odd and } 0 \leq i \leq n\}$.

7.2   Circuit and Function Terminology

In order to complete the discussions regarding symmetric functions, OB-DDs, and combinational circuits in the remainder of this chapter, it is necessary to define some terminology.

Definition 7.1 *The* cone of influence *of a line in a circuit consists of all the gates and primary inputs which feed the line either directly or indirectly.*

Definition 7.2 *Functional* decomposition, *as used here, means that some function $f(x_i, \ldots, x_j)$ is represented by a single switching variable in symbolic calculations. The complementary operation to decomposition is* composition, *where the single variable is replaced by the function that it represents.*

Definition 7.3 *A function $f(x_1, \ldots, x_n)$ is independent of variable x_i if:*

$$f(x_1, \ldots, x_n) = f(x_1, \ldots, x_i = \overline{v}, \ldots, x_n) = f(x_1, \ldots, x_i = v, \ldots, x_n). \quad (7.1)$$

Definition 7.4 *A function $f(x_1, \ldots, x_n)$ is dependent upon variable x_i if it is not independent of x_i.*

Definition 7.5 *An* interior *vertex in an OBDD is a vertex which is not a terminal vertex.*

Definition 7.6 *A function $f(x_1, \ldots, x_n)$ is* intrinsic *if it depends on all the variables $x_1, \ldots, x_n$.*

Definition 7.7 *Circuit* levelization *for combinational circuits is typically performed in the following manner.*

1. *Assign all gates fed by primary inputs only a level value of 1.*

2. *For each of the remaining gates in the circuit, once all its feeding gates' levels are known, assign its level to be $max(feeding\ gates'\ level) + 1$.*

In this way, no gate is fed by another gate whose level is equal to or greater than its level. We can also think of the gates which have an equivalent level assignment, i, as residing in level i.

Definition 7.8 *A function $f(x_1, \ldots, x_n)$ is* negative *in x_i if there exists a sum of products representation of $f(x_1, \ldots, x_n)$ such that only $\overline{x_i}$ (and no x_i) appears.*

Definition 7.9 *A* non-controlling *value at the input of a logic gate is a value that does not automatically imply the value at the output of the gate (under fault free conditions) regardless of the other input values.*

Definition 7.10 *A gate-level representation of a circuit is* planar *if it can be drawn in such a way that no two lines in the circuit intersect one another* [DEO74].

Definition 7.11 *A function $f(x_1, \ldots, x_n)$ is* positive *in x_i if there exists a sum of products representation of $f(x_1, \ldots, x_n)$ such that only x_i (and no $\overline{x_i}$) appears.*

Definition 7.12 *A* primary input *or PI is a line in a circuit that is not fed by any gate in the circuit. It is usually connected to some outside source of logic values.*

Definition 7.13 *A primary output or* PO *is a line in a circuit that does not feed any other gate in the circuit. It is usually connected to some outside point to which its values are passed.*

Definition 7.14 *A combinational circuit is* tree-structured *if no line in the circuit feeds more than one gate input.*

Definition 7.15 *A function $f(x_1, \ldots, x_n)$ is* unate *if it is either positive or negative in each of its switching variables. Functions which are not unate are called* nonunate.

7.3   The Symmetry Diagram

Shannon's symmetric function representation form is significant because it emphasizes one very important property of symmetric functions. Shannon showed with his "circuit for realizing the general symmetric function" that a simple expansion of a symmetric function is limited to $O(n^2)$ [SHAN38]. This simple expansion is very closely related to the concept of an OBDD.

OBDD size is held in check primarily by the mechanism of shared functional residues. By this we mean that no functional residue appears more than once in the correct OBDD representation of a function. Depending on the amount of residue sharing occurring within a function, the simple expansion of a function can range in size complexity from $O(n)$ to $O(2^n)$. If it can be shown that Shannon's "circuit" can always be constrained to obey the properties of OBDDs, then symmetric functions are guaranteed to exploit the residue-sharing property of OBDDs to a greater extent than almost any function imaginable. This constrainment indeed occurs, as will be shown.

Consider Figure 7.1. We refer to this diagram as the *symmetry diagram* of a general symmetric function of n variables. In the labeling of the vertices in the symmetry diagram, f_M is the symmetric function f with M-set M. $f_{M-k}(x_i, \ldots, x_n)$, denoted inside the vertices as $\sum k$, is the residue which exists after exactly k input variables (edges) have Boolean value = 1 (of the $i - 1$ that have been expanded). Stated more formally, this is

$$M - k = \{j \mid j + k = i \in M\}. \tag{7.2}$$

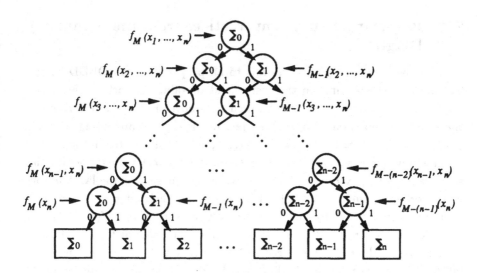

Figure 7.1: An illustration of the symmetry diagram for an n input symmetric function.

The symmetry diagram contains

$$\sum_{i=0}^{n} i = \frac{n(n+1)}{2} \tag{7.3}$$

interior vertices. Each of the $n + 1$ termini of the symmetry diagram represents one of the unique sums from $\{0, \ldots, n\}$. The Boolean value assigned to the terminus should agree with the functional result assigned to that sum. The terminus is assigned $v = 1$ if it represents the sum i and $i \in M$, else it represents a sum $j \notin M$ and the terminus is assigned $v = 0$. The symmetry diagram can represent all 2^{n+1} symmetric functions by assigning values from the set $\{0, 1\}$ to the $n+1$ termini. The symmetry diagram is functionally complete for symmetric functions because every possible residue of any symmetric function is represented within the diagram. The OBDD for any symmetric function of n variables cannot have more interior vertices than the symmetry diagram because the OBDD cannot contain more than every possible unique symmetric residue.

7.4 Removing Redundant Vertices from the Symmetry Diagram

The symmetry diagram conforms to all properties of an OBDD except 3.5 and 3.6. Depending on the assignment of values to each of the $n + 1$ termini, these two properties can be used to remove or consolidate vertices in the symmetry diagram. Property 3.5 removes a vertex representing x_i when $f(x_i = 0, x_{i+1}, \ldots, x_n) = f(x_i = 1, x_{i+1}, \ldots, x_n)$. As illustrated in Figure 7.2, when a contiguous set of j termini exists in the symmetry diagram with the same value, v ($\overline{v}$), exactly $j - 1$ interior vertices representing the last variable in the ordering (index n) must all have both of their edges directed to a terminus with value v ($\overline{v}$), and for these vertices $f(x_n = 0) = f(x_n = 1) = v$ ($\overline{v}$). Because the result of f is independent of the value of the variables in such residues, these $j - 1$ vertices should be removed, as well as the $j - 2$ interior vertices at index $n - 1$ which only have paths to the v ($\overline{v}$) termini, ..., etc., until a single vertex is removed. This process results in the removal of

$$\sum_{i=0}^{j-1} i \tag{7.4}$$

vertices. We will perform this operation for several classes of symmetric functions below.

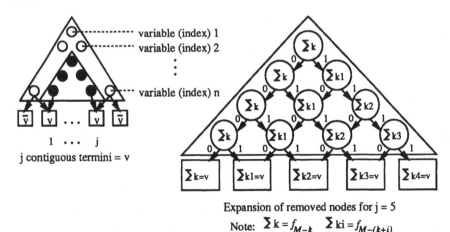

Expansion of removed nodes for j = 5

Note: $\sum k = f_{M-k}$ $\sum ki = f_{M-(k+i)}$

Figure 7.2: An illustration of the assignment of v (or $\overline{v}$) to j contiguous termini.

Before we can begin with the analysis of symmetric functions, we must prove the following statement.

Statement 7.1 *An OBDD which represents an intrinsic function $f(x_1, \ldots, x_n)$ of n variables must contain at least one interior vertex representing $x_i \ \forall \ i \in \{1, 2, \ldots, n\}$.*

Proof *Assume without loss of generality an OBDD representing the function $f(x_1, \ldots, x_n)$ which does not contain any vertices labeled x_i. By the definition of OBDDs appearing in [BRYA86], the value of any set of input conditions can be evaluated by tracing paths through the OBDD. Because no paths contain a vertex representing x_i, the value of f can always be determined regardless of the value of x_i. Thus, f is independent of x_i. This is in contradiction to our assumption that f is intrinsic, so the OBDD must contain at least one vertex representing x_i. This argument holds true $\forall \ i \in \{1, 2, \ldots, n\}$.* □

7.5 Derivation of OBDD Size Equations

As outlined in Section 7.4, we will proceed to derive equations which bound the sizes of OBDDs for specific classes of symmetric functions by using Property 3.5. In a later section, we will show that after having enforced Property 3.5 over the symmetry diagram in this way, each of the remaining vertices represent unique residues and must therefore remain in the graphs. Because the graphs are shown to obey all six OBDD properties, they are in fact OBDDs, and the equations thus give exact sizes of the representations for their respective classes.

7.5.1 Trivial Symmetric Functions

This symmetry diagram has all termini assigned v or all assigned $\bar{v}$. It represents the functions $f=0$ or $f=1$. Thus, from the discussion above, all the interior vertices in the symmetry diagram are redundant, and we have:

Statement 7.2 *Exactly 0 interior vertices exist in an OBDD representing an n input f=0 or f=1 function.*

7.5.2 Symmetric Functions with n Contiguous Termini $= v \ (\bar{v})$

This symmetry diagram has either all but the first or all but the last termini assigned the same value $v \ (\bar{v})$. By setting v to be 0 and then 1, we generate diagrams representing the common Boolean primitive functions NOT, AND, OR, NAND, and NOR. Using our removal process, we can subtract the redundant interior vertices from the total interior vertices in Equation (7.3) to obtain

Statement 7.3 *There are*

$$\sum_{i=0}^{n} i - \sum_{i=0}^{n-1} i = n \tag{7.5}$$

interior vertices in an OBDD representing an n input Boolean NOT, AND, OR, NAND, or NOR function (NOT is defined only for n=1).

Because an OBDD for an n input symmetric function must contain at least n interior vertices by Statement 7.1, this is an exact size for these functions.

7.5.3   Symmetric Functions of the Form $S_{\geq k}$, $S_{<k}$

The class of symmetric functions, $S_{\geq k}$, includes any M-set of contiguous integers $\{k, \ldots, n\}$ for any $0 \leq k \leq n$. Examples of M-sets in this class are $M = \{0, \ldots, n\}, \{1, \ldots, n\}, \{2, \ldots, n\}, \ldots, \{n-2, \ldots, n\}, \{n-1, n\}$, or $\{n\}$ as k varies from 0 to n, respectively. $S_{<k}$ is simply the complementary function to $S_{\geq k}$, and its M-set is the set complement of the M-set of $S_{\geq k}$. For the remainder of this discussion, $S_{\geq k}$ will be discussed, but the results obtained are equally valid for the case of $S_{<k}$. In terms of the symmetry diagram, these functions are characterized by one set of contiguous assignments, v, for sums $\{0, \ldots, k-1\}$, followed by another set of contiguous assignments, $\overline{v}$, for the remaining sums, $\{k, \ldots, n\}$. This defines two sets of vertices that must be removed from the symmetry diagram as in Figure 7.2, one set having k contiguous value assignments to one terminus value, while the other set has $n - (k-1) = n - k + 1$ contiguous assignments to the other terminus value. This leaves a maximum of:

$$\sum_{i=0}^{n} i - \sum_{i=0}^{k-1} i - \sum_{i=0}^{n-k} i = nk + k - k^2 \tag{7.6}$$

interior vertices for $S_{\geq k}$, $S_{<k}$, $0 < k \leq n$. The second summation is undefined for $k = 0$, but this k value is covered by Statement 7.2 for $S_{\geq k}$ ($f = 1$) as well as $S_{<k}$ ($f = 0$). This upper bound is also the exact equation as will be

shown by the uniqueness argument of Section 7.6. That argument proves that Properties 3.5 and 3.6 are met by each vertex remaining in the above graph.

Equation (7.6) is plotted in Figure 7.3 for n arbitrarily set at 80. From this plot it is apparent that the majority (over two thirds) of the functional instances have OBDDS with sizes that are within one half of the size of the worst case function for this class.

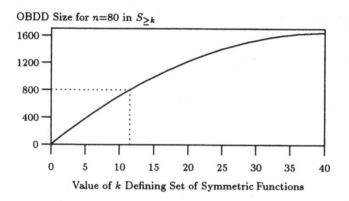

Figure 7.3: A plot showing OBDD sizes of $S_{\geq k}$ ($S_{<k}$) symmetric functions for $n = 80$.

7.5.4 Symmetric Functions of the Form $S_{=k}$, $S_{\neq k}$

The class of symmetric functions, $S_{=k}$, includes any M-set containing a single integer, k, $0 \leq k \leq n$, e. g. $M = \{0\}, \{1\}, \{2\}, \ldots, \{n-2\}, \{n-1\}$, or $\{n\}$ as k varies from 0 to n respectively. $S_{\neq k}$ is simply the complementary function to $S_{=k}$, and its M-set is the set complement of the M-set of $S_{=k}$. In terms of the symmetry diagram, these functions are characterized by one set of contiguous assignments, $\bar{v}$, for sums $\{0, \ldots, k-1\}$, a single, isolated v for sum k, followed by another set of contiguous assignments, $\bar{v}$, for the remaining sums, $\{k+1, \ldots, n\}$. This defines two sets of vertices that must be removed, one set having k contiguous value assignments to the termini, while the other has $n - k$ contiguous assignments. This leaves a maximum of:

$$\sum_{i=0}^{n} i - \sum_{i=0}^{k-1} i - \sum_{i=0}^{n-k-1} i = nk + n - k^2 \tag{7.7}$$

interior vertices for $S_{=k}$, $S_{\neq k}$, $0 < k < n$. For $k = 0$ and $k = n$, one of the above summations is undefined, but these k values are covered by Statement 7.3 for cases $S_{=k}$, $k = 0$ (f=NOR); $S_{=k}$, $k = n$ (f=AND); $S_{\neq k}$, $k = n$ (f=NAND); and $S_{\neq k}$, $k = 0$ (f=OR). The uniqueness argument will prove this upper bound to be exact.

Equation (7.7) is again plotted in Figure 7.4 for $n = 80$. Again, a large proportion of the total functions in this class fall within one half the size of the worst case function.

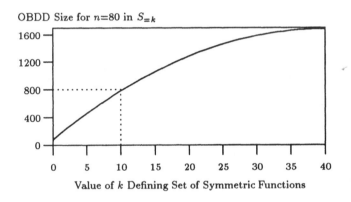

OBDD Size for n=80 in $S_{=k}$

Value of k Defining Set of Symmetric Functions

Figure 7.4: A plot showing OBDD sizes of $S_{=k}$ ($S_{\neq k}$) symmetric functions for $n = 80$.

7.5.5 Symmetric Functions with M-sets Integrally Divisible by a Constant c (not Divisible by c)

The M-set of an n input symmetric function is considered integrally divisible by a constant, c, if M is the set of integers from $0 \leq i \leq n$ such that i is divisible by c. We will denote this function as $c \mid M$. Assuming that $n > 0$, each of these M-sets contains the integer 0, and every cth integer after that up to n, including n if it is a multiple of c, i. e. $M = \{0, c, 2c, \ldots, jc\}$. For $c = 2$, $M = \{i \mid i \text{ is an even integer, and } 0 \leq i \leq n\}$, and represents the symmetric function NXOR (the complement of the exclusive-or function). The complementary function, M not divisible by c, is represented by M-set $M = \{i \mid i \text{ is an odd integer and } 0 \leq i \leq n\}$, and is the XOR function for $c = 2$. This function will be denoted as $c \nmid M$. To calculate a tight upper bound on the size of the OBDD, |OBDD|, for this class of functions, it is necessary to apply Property 3.6 of OBDDs to remove isomorphic residues from the symmetry diagram. This is done via the relation:

Statement 7.4 *For* $c | M$, $f_M(x_j, \ldots, x_n) = f_{M-c}(x_j, \ldots, x_n)$.

Proof *Let* f *be a symmetric function where* $c | M$. *Let* $f_{M-c}(x_j, \ldots, x_n)$ *and* $f_M(x_j, \ldots, x_n)$ *be two residues rooted at index* j *of that diagram. Assume that some set of values from* $\{0, 1\}$ *is assigned to* $\{x_j, \ldots, x_n\}$ *in both* f_{M-c} *and* f_M, *and let* w *be the number of 1 values assigned. From the definitions of these residues in Equation (7.2), the path from* f_M *must terminate at terminus* $\sum w$, *while the path from* f_{M-c} *must terminate at* $\sum w + c$, *where* $0 \le w, w + c \le n$. *But if* $c | w$, *then* $c | (w + c)$, *and if* $c \nmid w$, *then* $c \nmid (w + c)$, *so the value,* $r_i = v$, *returned by each function must be identical for all possible assignments to* $\{x_j, \ldots, x_n\}$, *and* $f_M(x_j, \ldots, x_n) = f_{M-c}(x_j, \ldots, x_n)$. $\square$

Statement 7.5 *In the expansion of a symmetric function (OBDD) where* $c | M$, *there are at most* c *unique residues (vertices) per variable (index).*

Applying Statement 7.5 to the symmetry diagram where $c \ge 2$ produces an acyclic digraph which has i vertices for each variable x_i, $0 \le i \le c$, then c vertices for each of the remaining $n - c$ variables. There are $c - 1$ termini, representing sums 1 through $c - 1$, being assigned the same $v = 0$ value, so that redundant subdiagram removal can be applied to those termini. This leaves:

$$\sum_{i=0}^{c} i + c(n - c) - \sum_{i=0}^{c-2} i = c(n - c + 2) - 1 \tag{7.8}$$

interior vertices for $c | M$ ($c \ge 2$) which for $c = 2$, gives $2n - 1$ interior vertices for XOR, NXOR. This equation is plotted for $n = 80$ as in the previous functional classes.

7.6 Uniqueness Argument

Each of Equations (7.6), (7.7), and (7.8) were derived as bounds. All of these equations will now be shown to be exact, rather than just bounds.

All functions considered in the preceding section were developed from symmetry diagrams with 1, 2, or 3 areas satisfying equation (7.4) being removed from the bottom (index n) of the graph upwards. Although the number

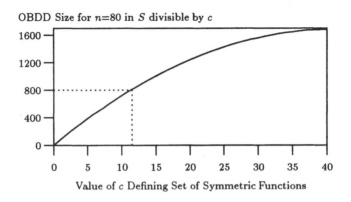

OBDD Size for $n=80$ in S divisible by c

Value of c Defining Set of Symmetric Functions

Figure 7.5: A plot showing OBDD sizes of $c\,|\,M$ $(c\,\!\!\!\!/M)$ symmetric functions for $n = 80$.

of areas removed varies for each case, the logic justifying the uniqueness of the remaining vertices is identical for all, except where noted. Consider any one of the graphs after removal of the redundant vertices. Clearly, each interior vertex remaining in the graph either does not border an area of removed vertices and has both edges to a different interior vertex (0-edge to $\sum k$ and 1-edge to $\sum k + 1$) as in Figure 7.1, or the vertex borders a removed area and it has one of its edges to an interior vertex and its other edge to the terminus for the removed area, as shown in Figure 7.6. Figure 7.6 is adequate to illustrate all of the symmetric classes considered, although all three removed areas exist in only one of those classes. Regardless of the case considered, none of the interior vertices remaining in the graph has both edges to the same vertex, so that if each interior vertex can be shown to represent a unique residue (to satisfy Property 3.6), then Property 3.5 must be satisfied for every interior vertex remaining in the graph for all cases. (Even the graph representing functions with $c\,|\,M$, $c \geq 2$, and $c\,|\sum k + 1$ has $\sum k$ and $\sum k + 1$ represented by different vertices).

Property 3.6 will be shown to hold by showing that the residues at index n are unique, and that equivalent residues at any index would require nonunique, equivalent residues at index n. For each equation developed in earlier sections, the one or two vertices remaining at index n represent unique functions (residues). This can be verified from the definitions of the functions for Equations (7.6), (7.7), and (7.8). In all cases where two vertices survive at index n, one vertex represents $f(x_n) = x_n$ while the other repre-

Figure 7.6: An illustration of the boundary vertices at index i of the symmetry diagram.

sents $f(x_n) = \overline{x_n}$, as can be verified by the v and $\overline{v}$ values at the termini to which the edges of each vertex point. Assume that two residues f_{v_1} and f_{v_2} represented by subgraphs rooted at vertices v_1 and v_2, $v_1 \neq v_2$, are equivalent functions in the remaining graph. Then each residue of f_{v_1} must equal the corresponding residue of f_{v_2}. Each function must be entirely replicated for all of these equivalent residues, or they must share some equivalent residue in the graph. No vertex which remains in any of the graphs has two 0-edges (or two 1-edges) incoming to it, so no vertex in the graph represents a shared residue. Because all remaining vertices have a path to both v and $\overline{v}$ (the ones which did not were removed), the vertices all represent residues of all $n - i$ remaining variables at index $i + 1$, and each must have some residue at index n. These must be equivalent, replicated residues at index n, so there must exist two vertices representing $f(x_n) = x_n$ or two vertices representing $f(x_n) = \overline{x_n}$. But this is not the case for any of the graphs developed. Thus, each vertex remaining in any of the graphs satisfies Property 3.5 and Property 3.6 of an OBDD, and the Equations (7.6), (7.7), and (7.8) are exact equations, as well as bounds.

None of the developments or arguments herein reflects any ordering dependencies, and these equations are valid for all $n!$ orderings of the input variables. Also, the results of Statement 7.3 and Equations (7.6) and (7.7) were previously derived by the authors using arguments which concentrate on the generation of the vertices that must remain in the graph, and equivalent equations were obtained [BUTL89].

7.7   OBDDs for Tree Circuits

The results of Statement 7.3 have implications beyond simply OBDD representations for symmetric functions. With some extensions, we can prove

OBDD sizes for certain circuit structures as well. We begin with some supporting statements.

Statement 7.6 *For any of the common primitive gate types (AND, NAND, OR, NOR, XOR, XNOR), if all the n inputs to a gate g realize functions of disjoint sets of switching variables and all those functions are dependent upon all of the switching variables in their respective cones of influence, then the output function of g, g(X), will be dependent upon all of the variables in its cone of influence.*

Proof *Select an arbitrary input from the set of n inputs to gate g. Call that input x_i. Place non-controlling values on each of the remaining $n-1$ inputs (either value for gate types XOR and XNOR). Trivially, these inputs can be justified because each of the $n-1$ input functions are disjoint and non-constant by assumption. Now, the output of gate g is determined solely by the value at x_i, and any of the switching upon which the function at line x_i depends can affect the output of gate g since the function at x_i is dependent upon all those variables. So, g(X) must depend upon at least all the switching variables upon which x_i depends. This argument can be repeated for each of the n inputs. So, gate g must realize a function which depends upon all of the switching variables upon which its inputs depend.* □

Now consider a single output combinational tree-structured circuit C built with primitives from the set {NOT, AND, OR, NAND, NOR} and realizing the function $f(x_1, \ldots, x_n)$.

Statement 7.7 *The function $f(x_1, \ldots, x_n)$ is dependent upon all of its switching variables.*

Proof *Let g be an arbitrary gate in the circuit realizing the function $f(x_1, \ldots, x_n)$. Then, by the definition of tree-structured circuits, each input of g realizes a function of some subset H_i of the circuit inputs $x_1, \ldots, x_n$. Moreover, for $i \neq j$, $H_i \cap H_j = \emptyset$. By Statement 7.6, the output function of a primitive gate fed by lines realizing disjoint input functions will be dependent upon all the variables in the functions of its inputs. This argument is true for any gate in the circuit g, so specifically it is true for the last gate in the circuit. The last gate in the circuit realizes the output function $f(x_1, \ldots, x_n)$. Thus $f(x_1, \ldots, x_n)$ is dependent upon all of the circuit inputs $x_1, \ldots, x_n$.* □

Statement 7.8 *For any n input combinational circuit which meets the criteria set forth for circuit C, there always exists at least one variable ordering such that the OBDD representing the function (using that ordering) realized at the output has exactly n interior vertices. Because no smaller OBDD is possible, this is an optimal ordering.*

Proof *Construct a planar representation of the circuit. Select the natural variable ordering that occurs from reading the PIs from top to bottom as they occur in the planar circuit representation. Note that this is one possible ordering resulting from the ordering heuristics presented in [FUJI88]. Starting at the PIs, construct OBDDs for all gates which are fed only by PIs. Each of these gates will have as many interior vertices in their representations as they have inputs from the result of Statement 7.3.*

Now assume for a moment that we decompose the functions that we have just generated into a new set of variables and generate the OBDDs for all of the gates in level 2. Again, due to the symmetric nature of these gates, we are guaranteed to have interior vertices which number as the inputs to the respective gates. If we were to now compose back the decomposed variables, we could use Bryant's simple composition procedure [BRYA86]. The simplified composition consists of replacing the decomposed variables with copies of the functions they represent. We can perform this operation because our natural ordering guarantees that the indices of the variables will always meet Bryant's criterion for performing this simple "composition by substitution" method. Because the OBDDs for the PI-only gates had interior vertices which numbered as their input sets, the second set of gates will now have OBDDs whose interior vertices number as the total number of PIs feeding them directly or indirectly. These arguments can be extended for each level of gates in the circuit until we finally reach the primary output of the circuit. Thus, the OBDD representing the output function will have exactly n interior vertices.

The ordering is optimal because by Statement 7.7 the function realized at the output must be dependent upon all n of its input variables. This requires at least one vertex of each index in the OBDD representation by Statement 7.1. Thus, the smallest the OBDD can be is to contain n interior vertices, and any OBDD of this size is optimal. □

Extension of the results of Statement 7.8 to encompass the inclusion of the XOR and XNOR primitives is difficult because they are nonunate functions. For any arbitrary tree circuit composed of any combination of gates from the set {AND, NAND, OR, NOR, NOT, XOR, XNOR}, it is conjectured that no closed form result to predict OBDD size at the output is possible.

Now assume a planar representation of a combinational circuit meeting the criteria set forth for circuit C. Form subsets of PIs called G_1 through G_k where two PIs are in the same subset if they both directly feed the same gate. Let $|G_i|$ denote the cardinality of subset G_i. This leads us to our next statement.

Statement 7.9 *Assume that we have a combinational circuit which meets the criteria of circuit C and that we have formed the set of subsets $\{G_1, \ldots, G_k\}$ as stated above. Then the number of optimal variable orderings for an OBDD representation of the function realized at the output of the circuit is not less than $|G_1|! \bullet \cdots \bullet |G_k|!$.*

Proof *The proof of Statement 7.8 was founded on the construction of a planar representation of the circuit. If we bound the number of legitimate distinct planar representations, we have then bounded the number of optimal orderings of the switching variables. Certainly planarity will not be disturbed if we rearrange the inputs within their separate sets. There are $|G_1|! \bullet \cdots \bullet |G_k|!$ total distinct permutations of the variables within their groups. Thus, there are not less than $|G_1|! \bullet \cdots \bullet |G_k|!$ optimal orderings of the switching variables.* □

7.8   OBDD Size Summary

Exact equations were derived which give the sizes in vertices of OBDD representations of the simplest classes of symmetric functions. Because most of this research has concerned the application of OBDDs to gate level circuits to gain useful design and test information, it is important to have explored the limits of the representation. Often, such circuits contain thousands to tens of thousands of gates through which functions must be calculated to reach a remote site or sites where the results are needed. Many of the remote sites in VLSI are functions of 60 to 100 inputs. If one optimistically assumes that the average such function will take only as much space as the average symmetric function from the classes studied herein, then a typical OBDD for such a function would require thousands of vertices. An OBDD as specified in [BRYA86] contains six fields which would occupy at least 15 bytes in any common computer language. Assuming only 1000 sites requiring functional representations have moderate to high numbers of inputs, we would anticipate that at least 15 megabytes of memory would be required to represent only the functional information at those sites. The use of residue sharing could reduce this memory requirement somewhat [MADR88], [BRAC90], [MINA90], but the extreme optimism of the other assumptions in this argument leads the authors to believe that the lower bound in this example is quite realistic. Thus, a straightforward approach of applying OBDDs to very large circuits is somewhat questionable.

Experimental work with the combinational benchmark circuits [BRGL85] detailed later in this monograph has shown that OBDD sizes often grow intractable in practice. Many would argue that the benchmark circuits are aberrant and that nominal behavior would be much more acceptable. The work summarized in this chapter leads the authors to conjecture that this is not the case.

Chapter 8

Difference Propagation

Topological circuit information can often be derived from algorithms with computational complexities linear in some size parameter such as gate count or input count. Functional information on the other hand, such as the *syndrome* (the proportion of ones in the Karnaugh map of the function, [SAVI80]) of a circuit line is often much more costly to obtain. While very fast fault simulators exist, e. g. [WAIC85], the sheer size of many circuits makes the exhaustive simulation of large fault sets impractical. As we saw in Section 4.3, the use of Boolean functional techniques, on the other hand, has shown promise for applications where a great deal of information is needed, such as redundancy proving [GAED88], [STAN88]. We have adopted a similar function-based approach in this research.

8.1   The Development of Difference Propagation

Difference Propagation was originally developed primarily as an alternative for comparison to CATAPULT [GAED88]. It is a combinational test generator which uses OBDDs as defined by Bryant [BRYA86] as the vehicle for its functional evaluations. Unlike CATAPULT, Difference Propagation does not derive its observability information disjointly from the control information, thus eliminating the need for explicit use of the Boolean difference [AKER59].

Much like CATAPULT, Difference Propagation must calculate functional information through levels of combinational circuitry. The circuit preprocessing and the generation of good circuit or controllability information are essentially the same as in CATAPULT. However, the process of generating observability information is merged with the actual test generation work. This is accomplished by explicitly addressing the difference between the good and faulty machines, much like symbolic fault simulation [CHO89] and hence the name.

This idea stems from considering the mechanism by which faults become detectable. A logical fault (usually) causes some perturbation in the function realized at some point in the circuit. The change in the function at the input to some gate often changes the function realized at the output of that gate. If this gate output is the input to another gate, the perturbation will often cause yet another change at that second output and so on until a primary output is reached. Ideally then we would like to be able to represent only the perturbations to the logic functions. In the next section we present the mathematics to accomplish this task.

8.2   Deriving the Input-Output Relationships

In representing only the differences between the two machines during the fault calculations, it is necessary to find relationships which allow only the difference information to pass through gates in the circuit. The derivation of the input-output gate relationships used in Difference Propagation takes advantage of the ring-sum properties of the exclusive-OR operator (XOR, $\oplus$) over the Galois field $GF(2)$. The multiplication operator in $GF(2)$, $\bullet$, will be implied by the juxtaposition of two terms in the derivation. In order to facilitate the discussion, consider Figure 8.1. In this simple single AND gate circuit, we will define three functions at each line in the circuit. For $i \in \{A, B, C\}$,

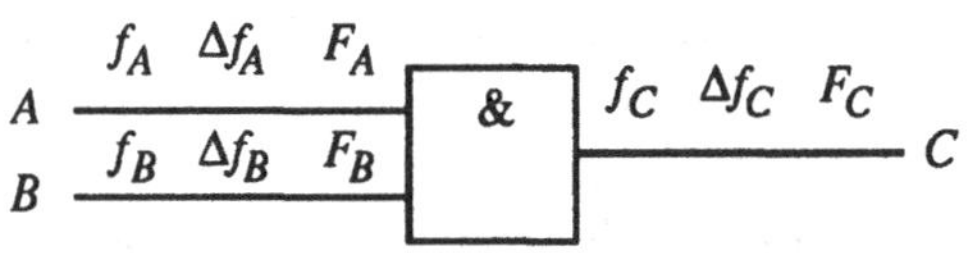

Figure 8.1: An example defining fault functions and difference functions.

Definition 8.1 f_i *is the Boolean function realized at line i under completely fault free conditions.*

Definition 8.2 F_i *is the Boolean function realized at line i under the influence of some particular fault.*

Now, in a manner very similar to the Boolean difference [AKER59],

Definition 8.3 Δf_i *is the Boolean function which describes all the conditions in which f_i and F_i differ. In other words,*

$$\Delta f_i = f_i \oplus F_i \tag{8.1}$$

What is needed now is the ability to calculate the difference function Δf_C in terms of the information we have at lines A and B. We know that

$$f_C = f_A f_B, \text{and} \tag{8.2}$$

$$F_C = F_A F_B. \tag{8.3}$$

From the the ring-sum property of $\oplus$ and Equation (8.1),

$$F_i = f_i \oplus \Delta f_i \ \forall \ i \in \{A, B, C\}. \tag{8.4}$$

Substituting the results of Equation (8.4) into Equation (8.3) yields

$$F_C = (f_A \oplus \Delta f_A)(f_B \oplus \Delta f_B). \tag{8.5}$$

But, by the distributivity of $\oplus$ over $\bullet$,

$$F_C = f_A f_B \oplus f_A \Delta f_B \oplus f_B \Delta f_A \oplus \Delta f_A \Delta f_B. \tag{8.6}$$

However, by substituting Equation (8.2) into the result of Equation (8.4) we get

$$F_C = f_A f_B \oplus \Delta f_C. \tag{8.7}$$

Now, we substitute this result into Equation (8.6) to obtain

$$f_A f_B \oplus \Delta f_C = f_A f_B \oplus f_A \Delta f_B \oplus f_B \Delta f_A \oplus \Delta f_A \Delta f_B \tag{8.8}$$

and use the ring's cancellation property to finally arrive at

$$\Delta f_C = f_A \Delta f_B \oplus f_B \Delta f_A \oplus \Delta f_A \Delta f_B. \tag{8.9}$$

Thus, we can express the difference function at the output of an AND gate in terms of only the good and difference functions at its inputs. Using similar derivations, the same is true for the remaining common combinational primitives. For the sake of brevity, we will not repeat the derivations, but

Gate type	Output Relation
AND/NAND	$\Delta f_C = f_A \Delta f_B \oplus f_B \Delta f_A \oplus \Delta f_A \Delta f_B$
OR/NOR	$\Delta f_C = \overline{f_A} \Delta f_B \oplus \overline{f_B} \Delta f_A \oplus \Delta f_A \Delta f_B$
XOR/XNOR	$\Delta f_C = \Delta f_A \oplus \Delta f_B$
INVERTER/BUFFER	$\Delta f_C = \Delta f_A$

Table 8.1: The formulas for calculating output difference functions in terms of input good and difference functions.

the results are summarized in Table 8.1. For each gate type (except the INVERTER and BUFFER) assume two inputs A and B and one output C and their good and difference functions as in the example above.

For gates with more than two inputs, analogous equations exist to formulate the output difference function. The problem with these equations is that all but one of the pairs, triplets, etc., of good and difference functions must be calculated. This phenomenon gives rise to operations whose number grows exponentially with the number of gate inputs. This exponential growth can be held in check by modeling an n input gate as $n - 1$ two input gates, and using the two input gate equations to calculate $n - 2$ intermediate results and 1 final result. The number of operations will then be less than in the general case. It is also worth noting that difference functions at random gates along the propagation paths are often identically 0, and all terms including those particular difference functions disappear from the calculations. Thus, in a manner analogous to *selective trace* from digital simulation theory [MENO65], calculations are only performed as long as difference information exists.

Finally, we should remark about the lack of dependence on fault type. The relationships in Table 8.1 did not in any way rely on particular properties of the fault model, for example fixed logical values at the fault site, etc. The only requirement of the fault is that its effect must be capable of being described logically. Many fault models in the literature have this property, and thus Difference Propagation can be used for a large variety of fault types above and beyond the stuck-at fault model.

8.3   The Difference Propagation Algorithm

A psuedo-code representation of the Difference Propagation algorithm appears in Figure 8.2. Difference Propagation first initializes the difference function(s) at the gate(s) feeding the fault site. It then uses the relationships shown in Table 8.1 to propagate difference functions towards the primary outputs in a manner much like conventional ATPG systems. In implementing the selective trace idea, we proceed through the circuit and calculate differences only along the paths where they exist. This is achieved through a gate marking scheme. At a marked gate, we calculate the difference function. If it is non-zero, then all gates fed by this gate are marked for further calculation. If the difference function is identically 0, then the difference along this path has "terminated", and no further differences exist between the good and faulty machines along this path.

This mark and calculate process continues until all marked gates up through and including the POs have been reached. The OR of the differences between the good and faulty functions at the POs is identically the complete test set for a fault. Any calculations needing to be performed on the test function, such as counting its minterms, can be performed once the complete test set has been calculated.

8.4   The Efficiency of Differences

At this point we should pause for a moment to motivate the use of differences. Because the fault function F_i completely describes the logical impact of a fault at line i, it can be used to represent the faulted information in Difference Propagation. Intuitively it was our belief, though, that the difference function Δf_i at line i would often be "smaller" than the fault function F_i at the same line. If that hypothesis were true, it would be less efficient to use the fault functions than the difference functions.

In order to facilitate a more rigorous comparison of the two techniques, Difference Propagation and Fault Function Propagation were both implemented [BUTL88]. The two techniques were compared to each other and to CATAPULT. The benchmark circuits used were four circuits from the standard combinational benchmark circuit set [BRGL85]. In order to compare to CATAPULT in its intended domain, the target fault sets were again the faults aborted by TOPS in nominal test generation runs (see Section 4.3.2) along with the testable faults at the same site. The results are shown in Table

```
function difference_propagation()
{
    for (each fault)
    {
        create_fault_site_difference_function();
        for (each gate fed by fault site)
            mark gate to be evaluated;
        for (each marked gate)
        {
            calculate_difference_function();
            if (this difference function != 0)
            {
                for (each gate fed by this gate)
                    mark gate to be evaluated;
            } /* end if (this difference ...) */
        } /* end for (each marked gate) */
        test_function = NIL;
        for (the primary outputs)
        {
            if (difference function at this PO exists)
            {
                if (test_function == NIL)
                    test_function = this_difference;
                else
                    test_function =
                        test_function OR this_difference;
            } /* end if (difference function ...) */
        } /* end for (the primary outputs) */
        if (test_function contains decomposed variables)
            compose_until_pi_variables_only();
        /* Perform desired calculations here. */
        unmark_all_marked_gates();
    } /* end for (each fault) */
    return;
}
```

Figure 8.2: A psuedo-code representation of the Difference Propagation algorithm.

8.2. The data were generated using programs written in C and compiled with the GNU gcc compiler with optimization and run on a SUN3 workstation under the UNIX programming environment. It should be noted that CATAPULT does not currently support bridging faults. Thus, the CPU time for CATAPULT for the circuit C432 is for the 12 stuck-at faults only.

ISCAS Circuit	Testable stuck-ats	Redund stuck-ats	Bridge Faults	CATAP CPU secs	Diff Prop CPU secs	Fault Func CPU secs
C432	6	6	1	9.0	24.1	25.0
C499	8	8	0	119.6	567.5	815.9
C1355	8	8	0	151.0	1562.4	984.1
C1908	2	2	0	156.7	157.6	189.7

Table 8.2: A comparison of total computation times for CATAPULT, Difference Propagation, and fault function propagation.

The data in Table 8.2 point out at least two interesting phenomena. First of all, our suspicion that using difference functions is more efficient than fault functions was confirmed by this and other experimentation. An examination of the total OBDD vertices required for test generation of each individual fault also indicated Difference Propagation to be the clear winner over fault function propagation. The other is that in these cases, CATAPULT takes less time to generate tests for single stuck-at faults than either Difference Propagation or fault function propagation. This can probably be attributed to the fact that CATAPULT is a concurrent test generation algorithm while the approach of both Difference Propagation and fault function propagation is fault serial.

There were two reasons for using Difference Propagation despite this disparaging observation. First, the overall approach of Difference Propagation is more amenable to the addition of non-single stuck-at faults than that of CATAPULT. Also, due to the large amount of decomposition and composition that must occur for large fault sets, CATAPULT's performance degrades to the point of being surpassed by Difference Propagation. This point will be discussed further in the next section.

8.5   Using Functional Decomposition

Functional decomposition can be useful in many aspects of Boolean functional manipulation. In the case of OBDDs, functional decomposition has the potential for speeding up calculations because subgraphs of many vertices are represented by a single vertex. Because the manipulation time of an OBDD is a function of its size [BRYA86], small OBDD size is a desirable feature. The efficiency gained through functional decomposition is not without cost, however. For the final results to be usable, the decomposed variables must often be composed out of the OBDDs to express them in terms of only input variables. Thus a tradeoff exists between decomposition and composition time, and the selection of decomposition sites is an important problem. In this section we describe various techniques for the selection of decomposition sites and discuss the results of experimentation with these techniques. Portions of this section are paraphrased from a paper co-written by the authors [HUNG89].

8.5.1   "Random" Decomposition

In Section 4.3.2 we noted that the use of the Boolean difference requires that circuit functions be expressed in terms of fanout stem variables as well as PI variables. The exact variables required by CATAPULT to calculate the Boolean difference depend on the locations and size of the fault set for which tests will be generated. Depending on the target fault set, this is in some sense a random selection of decomposition sites. Unfortunately, letting the target fault set guide the selection of decomposition sites rarely results in a set of decompositions which optimize the propagation of functional information. Furthermore, the larger the target set, the more decomposition is necessary. This increased amount of decomposition often has detrimental results. Thus, it seems that better heuristics are needed to utilize the notion of functional decomposition.

8.5.2   Threshold Decomposition

In Fujita's recent work on variable ordering in OBDDs, he proposes the idea of "threshold decomposition" [FUJI88]. Stated simply, threshold decomposition mandates the selection of an arbitrary OBDD size threshold. When the size of an OBDD at the output of a gate is above this threshold, decomposition automatically occurs at that point and propagation continues using the altered set of variables. This process of thresholding followed by decomposition is repeated as necessary until the primary outputs are reached, at which time composition occurs to obtain output OBDDs in terms of primary inputs only.

Threshold decomposition was implemented and several limitations were revealed. First of all, the arbitrary threshold is difficult to find, and seems to be circuit and function dependent. Secondly, if the threshold is set too low, far too much decomposition occurs, and the time invested in composition at the outputs outweighs the savings obtained during the forward propagation. If, on the other hand, the threshold is set too high, few decompositions occur and little time savings is realized. Furthermore, this again is a quasi-random approach, which uses little topological information from the circuit. Our findings indicated that the topological characteristics of the decomposition site are very important, as will be discussed in the next section.

8.5.3 Minimum Circuit Width Approach

Berman suggests the notion of using the width of a circuit as a guide to selecting the decomposition set [BERM88]. Berman's definition of circuit width is closely related to the graph theoretical concept of the minimal cut set [DEO74], and can be intuitively explained as follows. Assume a combinational circuit that has been levelized from primary inputs to primary outputs in the traditional manner (see Definition 7.7 in Section 7.2). The minimum width of a circuit as defined by Berman is then the cardinality of the smallest set of gates having equivalent levels, which can be seen as closely analogous to the definition of the minimal cut set of a graph. The minimal cut set of a graph is defined as the smallest set of edges whose removal disconnects the graph. Berman proposes that decomposition should be performed at the level of minimum width, but presents no experimental results in his work [BERM88].

In [HUNG89], the concept of minimum width was extended by slightly relaxing the level requirements. This idea is best illustrated with an example. The circuit in Figure 8.3 is a typical example to describe our algorithm in calculating the modified width of a circuit. At first, all the gates in the circuit are leveled (when counting levels, we neglect all the single input gates since the size of an OBDD at the output of a one input gate is exactly the same as the size of the OBDD at the input). If k is the number of levels in the circuit, then define an array of cut sets CUT[1] through CUT[k]. CUT[1] is defined as those gates with at least two inputs, all of which are PIs. In our example, CUT[1] consists of the gates $A1$-$A5$. At first glance, CUT[2] consists of only gates $B1$-$B4$. Because the output lines of gates $A1$-$A5$ feed gates in the third level as well as gates $B1$-$B4$, we modify CUT[2] to include gates $A1$-$A5$ and $B1$-$B4$. In this way, the CUT[3] consists of gates $C1$-$C5$, CUT[4] consists of gates $D1$, $D2$, and $C3$, CUT[5] consists of gates $D1$, $D2$, and gates $E1$-$E4$, and CUT[6] consists of only gates $F1$-$F5$. With these definitions,

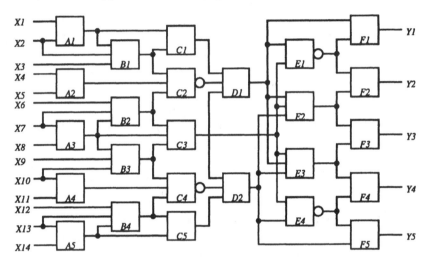

Figure 8.3: An example circuit to illustrate location of minimum width decomposition points.

it is clear that CUT[4], which consists of only 3 gates, is the best place to perform decompositions. Figure 8.4 presents a pseudo-code representation of this method for calculating modified circuit width as described above.

8.5.4   Empirical Comparison of Decomposition Techniques

We now present the results of an empirical comparison of these decomposition techniques. We used a subset of five of the combinational benchmark circuits [BRGL85]. The fault sets used were again the sets of "hard" single stuck-at faults which were "dropped" by TOPS with a low backtrack limit [KIRK87], along with the corresponding testable faults at the same sites, just as was done in the experimentation described in Section 8.4.

Table 8.3 describes the characteristics of the circuits and fault sets used in this study and details the performance of CATAPULT as a reference. The first column in the table gives the name of each circuit as it appears in the

```
function find_min_width()
{
    Put gates with 2 or more PI only inputs into CUT[1];
    k=1;
    for (all k)
    {
        for (each gate G in CUT[k])
        {
            for (each gate F that is driven by G)
            {
                if (level(F) - level(G) == 1)
                    put F into CUT[k+1];
                else
                    put G into CUT[k+1];
            }
        }
        if(|CUT[k+1]| == 0)
            return;
    }
    return;
}
```

Figure 8.4: A psuedo-code representation for function find_min_width().

benchmark data. Columns 2-4 give the numbers of primary inputs, gates, and
levels as determined by standard forward levelization, respectively. Column
five gives the total numbers of faults used in these comparisons for each of the
circuits. Column six gives the total CPU time for one test generation run.
The left half of column seven gives the size in vertices of the largest OBDD
created during the test generation process, and the right half gives the total
number of OBDD vertices created.

Tables 8.4 and 8.5 summarize our experimental results for the various
forms of Difference Propagation. The column definitions are identical to
those of Table 8.3. The columns labeled "Dec" refer to the total numbers
of functional decompositions that occurred at the minimum width regions.
The table shows that the addition of decomposition to Difference Propagation makes possible a 2 to 60 times speedup in CPU time. This reduction
can be easily understood by examining the maximum and total OBDD vertices used during processing, again because all the OBDD operations' time

complexities are functions of the operand OBDD sizes. Furthermore, OBDD sizes are a function of the ordering of their input variables (see Chapter 3), but no attempt was made to locate "good" variable orderings when gathering this data. The orderings used were those dictated by the ISCAS source information [BRGL85].

ISCAS Circuit	Total inputs	Total gates	Total levels	Faults in target set	CPU secs	Vertices max	Vertices total
C432	36	203	7	12	9	484	25797
C499	41	275	11	16	119	5090	302059
C1355	41	619	24	16	151	5090	383854
C1908	33	938	40	4	156	5555	292316
C3540	50	1741	47	4	Out of Memory		

CATAPULT

Table 8.3: A summary of benchmark circuit characteristics and CATAPULT's performance.

Ckt	Difference Propagation CPU secs	Vertices max	total	Difference Propagation with decomposition CPU secs	Vertices max	total	Dec
C432	25	494	59792	14	362	42635	9
C499	567	9483	784270	22	1811	43832	8
C1355	1562	9483	2211939	27	1811	51092	8
C1908	157	5555	292660	21	1416	40338	37
C3540	Out of Memory			71	11339	131938	87

Table 8.4: A comparison of Boolean functional test generation computation resources using Difference Propagation without and with functional decomposition.

Table 8.5 illustrates an interesting refinement to the decomposition technique. After reaching the modified minimum width set of gates, the sizes of the OBDDs at each of the lines on the outputs of these gates are examined. Decomposition is then performed only at those lines whose OBDD size is above a very low threshold (approximately 10 vertices). In this way, the notions of modified minimum width and thresholding are combined, and the results are further reductions in both space and time.

	Difference Propagation with decomposition and thresholding			
	CPU	Vertices		
Ckt	secs	max	total	Dec
C432	14	362	42635	9
C499	22	1811	43832	8
C1355	27	1811	51092	8
C1908	17	789	33639	23
C3540	58	7461	110389	56

Table 8.5: A summary of Boolean functional test generation computation resources using Difference Propagation with functional decomposition and thresholding.

It should be noted here that for the circuit C3540, two minimum width regions were selected for decomposition instead of just one. Much less improvement in performance was realized with only one region of decomposition. Thus, for circuits as large as or larger than the circuit C3540, it may be better to choose more than one low width set of gates to decompose, or to choose a combination of gates in several low width sets. A rigorous response to this question remains an interesting topic for future research.

Chapter 9

Fault Model Behavior

One aspect of paramount importance in deterministic testing is the performance of the fault models underlying the process. Recent research has shown that the fabrication process of a circuit is certainly an issue when measuring fault model performance [SHEN85]. However, it is also useful and informative to consider the functional limitations of fault models relative to the circuits themselves and independently of the technology chosen to realize them. Exhaustive simulation or simulation of particular test sets is one possible method that can be used to attack this sort of problem [HUGH86], [MILL88], [MILL89]. However, this approach is limited to relatively small samples of test sets due to otherwise exorbitant computation time requirements.

In this chapter we discuss a new method for gathering this type of information and we present results of experiments with a set of benchmark circuits, some of which are from the combinational benchmark set [BRGL85]. In increasing order of size, our circuit set is C17, a fulladder circuit, C95, the 74LS181 arithmetic logic unit (ALU), C432, C499, C1355, and C1908.

This chapter is based on the paper "The Influences of Fault Type and Topology on Fault Model Performance and the Implications to Test and Testable Design", which first appeared in the *Proceedings of the ACM/IEEE 27th Design Automation Conference*, Orlando, Florida, June 24-28, 1990, pp. 673-678.

9.1 Selection of Fault Models and Fault Sets

This work is aimed at the two most widely studied fault models. These are the classical stuck-at fault model and the bridging fault model. Each of these two models are described in further detail in Section 2.2.3. The characteristics and assumptions pertinent to this particular study of the models are discussed briefly below.

9.1.1   Stuck-at Fault Sets

The stuck-at fault sets chosen for this work are based on the well-known concept of checkpoint stuck-at faults [BOSS71] (see Section 2.2.3). Checkpoint faults have been chosen because of their wide acceptance in the literature and industry as good targets for test generation. The checkpoint fault sets are further reduced by applying the notion of fault equivalence [MCCL71] at gate inputs. This is done to make the number of representatives from each fault class be as small as possible. Through minimization of faults in each fault class, we have been able to obtain data that was as close to being independent of the circuit realization as possible. Also, duplication of test generation effort has been avoided in many cases.

9.1.2   Bridging Fault Sets

The bridging fault model has experienced a growth in popularity. Research has shown that for the currently popular MOS technologies, the bridging fault model may be the dominant model for actual failure mechanisms [GALI80]. It is thus unfortunate that few combinational test generators today explicitly address bridging faults.

The bridging fault model employed in this study will be similar to those used by [MEI74] and [ABRA83]. We will utilize BFs between two lines only in accordance with most researchers' assumption that three or more lines bridged together is a fairly unlikely occurrence. Unlike [ABRA83], we will use both AND BFs and OR BFs. While the actual behavior of a BF is certainly technology dependent, it is commonly understood that zero-dominant logic often gives rise to BFs behaving as a wired-AND, and one-dominant logic wired-ORs. In modeling both types of BFs, we will be able to see the implications, if any, of one logical scheme over the other.

We have chosen to address only non-feedback bridging faults (NFBFs) for two reasons. First, Millman and McCluskey showed empirically that feedback bridging faults (FBFs) are much more likely to be detected by high coverage stuck-at test sets than are non-feedback bridging faults [MILL88]. Secondly, our analysis technique is unidimensional, purely functional, and based on a gate-level circuit representation. Thus, accurate modeling of induced sequentiality, intermediate logic values, and other non-binary behavior is not possible. Finally, it should be mentioned that our logical modeling of bridging faults implicitly assumes a large driving capability of individual gates. If the gate that drives the controlling value of the bridging fault is bridged to a fanout stem with many branches, voltages outside the logic thresholds could result if fanout limits are low.

The sets of NFBFs used here could be best described as "potentially detectable" NFBFs. That is because all trivially undetectable NFBFs (such as the AND NFBF between two inputs of the same AND gate) were screened out during the fault set generation process.

While the four smallest circuits had rather large NFBF sets, the entire set was still used for each. For C432 and the other three larger circuits, the sizes of the NFBF sets dictated another approach. Since no analog to checkpoint faults exists for BFs, it was decided to select faults at random from the set of all the NFBFs. However, it is clearly true that due to layout concerns, not all NFBF are equally likely to occur.

The lack of available layout information for the benchmark circuits led us to an approximate solution. The approach is based on estimates of the line distances in any potential layout. Each gate is assigned an X coordinate based on its distance in levels from the primary inputs. The Y coordinates of the gates are calculated as follows. First the n PIs are assigned Y coordinates 0 through $(n-1)$ in the order in which they were given in the original benchmark data [BRGL85]. This assumes that the PIs were stated in a meaningful order. Our work with variable ordering in OBDDs indicates that this assumption is probably valid. Then, level by level, each gate is assigned a Y coordinate in its level based on the average of the Y coordinates of all the gates feeding it. In this way, we consider the aggregate of all possible layouts for that PI ordering without having to select one and adhere to it.

Once these approximate X and Y coordinates are generated for each gate, the distances between the two lines of each NFBF are calculated using the standard distance equation for a 2-dimensional Cartesian coordinate system. These distances are normalized to the largest distance of all the potentially detectable NFBFs. Faults are then selected randomly assuming that these normalized distances, z, are distributed exponentially, that is with probability distribution function

$$f(z) = \frac{1}{\theta}e^{-z/\theta}. \tag{9.1}$$

Since the fault sets are selected randomly, we can still expect their behavior to be representative of the entire set, but at the same time be based on actual probabilities of occurrence. The value of θ was adjusted to facilitate fault sets of reasonable sizes (≈ 1000 faults).

9.2   Fault Behavior Results and Analysis

In this section we present statistics which were gathered using Difference Propagation on our benchmark set. The data will be presented graphically with discussion of several interesting inferences which can be drawn from them.

The complete set of analyses were carried out on each circuit but many of the results were similar among the circuits. In an effort to avoid redundancy, samples of the analyses will be stated for certain fault types and circuit instances with the understanding that the results are typical of the entire benchmark set. These results will be meaningful inasmuch as the investigated circuits [BRGL85] are representative of typical combinational designs.

9.2.1   Stuck-at Fault Behavior

We begin by examining the detection probability histograms for our example circuits. Several researchers have investigated these profiles through approximate means, an example of which is [AGRA85]. Since we generate the complete test set for each fault, we can calculate the exact detection probability for each fault as well. Instead of reporting raw numbers of faults, we normalized the fault counts to the fault set size. An example of a detectability profile is shown in Figure 9.1 for the ISCAS circuit C95 and the 74ls181 ALU.

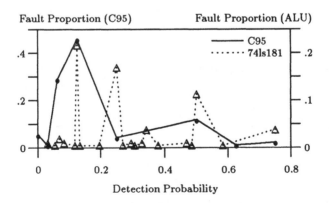

Figure 9.1: The stuck-at fault detection probability histograms for circuits C95 and 74ls181.

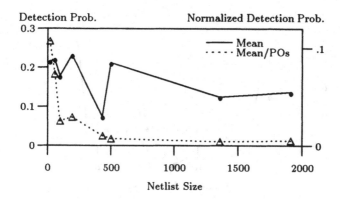

Figure 9.2: The trends of mean stuck-at fault detection probabilities and ratioed detection probabilities for the example circuits.

The set of all these profiles leads to the hypothesis that detection probabilities (and thus testability) are decreasing with increasing circuit size. This observation is corroborated by Figure 9.2. The solid line represents the overall mean detectability of detectable faults versus netlist size for the circuit set. This plot does not reveal a true trend of detectability. However, it was noted that the increases in numbers of POs are not proportional to the increases in PI counts in these circuits. The dotted line represents the same trend when the detectable faults' overall mean detectability is normalized to the number of POs in the circuit. Here we can see the decrease in detectability with increasing circuit size.

This plot is particularly interesting in light of the inclusion of both the circuits C499 and C1355. Since C1355 is identical to C499 except with exclusive-ORs expanded into their four-NAND equivalents, one would expect that their detectability profiles would be the same. It can be seen, though, that the detectability still decreases with the added circuitry. The desirability of minimal designs due to testability concerns is thus established.

The obvious inter-relationships between controllability, observability, and testability inspired a study of the relationship between a fault's detectability and its PI and PO proximity. The discovery of a general relationship between the topology of a fault and its testability would be useful for several reasons. For example, such information would provide the test engineer with some a priori knowledge about where most test generation effort would be expended. It would also provide hints as to how to best modify circuits when adding

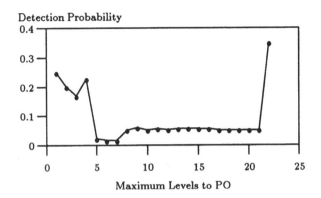

Figure 9.3: A plot of mean stuck-at fault detectability versus maximum distance to POs for circuit C1355.

design for testability (DFT) hardware. Should the emphasis be placed on additional control lines or observation points?

One such testability curve is reproduced in Figure 9.3. The study resulted in PO distance curves reminiscent of the famous "bathtub curve" of reliability theory. The curves support the intuitive notion that both highly controllable and highly observable faults are more easily detected than are those near the center of the circuit. They also suggest that most DFT modifications should target the region just upstream of the circuit outputs where reconverging signals render faults difficult to detect [AGRA90]. What was less apparent at the outset is that detectability seems more closely correlated with observability than with controllability. Plots of detectability versus PI distance often were much more random than detectability-PO distance plots. This result coupled with that of Figure 9.2 suggests that detectability is best increased through enhanced observability, which contradicts the result given by Fujiwara [FUJI90]. This question has been studied further using our techniques. A discussion of this investigation is given in Chapter 10.

Another interesting observation which will not be quantitatively supported relates more to testing than design. The POs fed by a fault site were counted and compared to the number of POs at which the fault was observable. These numbers are almost always the same. This simple observation leads one to believe that the testing heuristic of justification to the closest PO will nearly always result in test effort reduction [ROSS89]. It is also further evidence that PO counts should be maximized for improved testability.

The last statistics for stuck-at faults that we will cite here relate to the syndromes of fault sites mentioned earlier. It is a trivial observation that to successfully test a fault, the controllability and observability functions must be satisfied simultaneously. In the case of a stuck-at-0 fault, certainly it is not possible for the size of the test set to be larger than the minterms in the controllability function. This is true because fault excitation requires a logic one on the faulty line. An equivalent statement would then be that the detectability of the fault cannot exceed the syndrome of the corresponding line. Thus, the syndrome of a line represents an upper bound on its detectability. For stuck-at-1 faults, the complement of the syndrome is the upper bound.

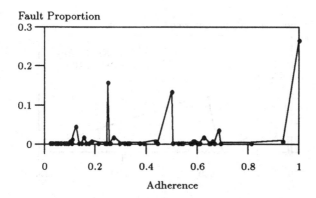

Figure 9.4: The stuck-at fault adherence histogram for circuit 74ls181.

Let δ_i be the detectability of some fault, i, and v_i be its detectability upper bound. Then, we define the *adherence* of the fault, α_i as

$$\alpha_i = \delta_i / v_i. \tag{9.2}$$

The adherence is simply the proportion of minterms exciting the fault which turn out to be tests. It is in some sense a measure of the usefulness of approximating detectabilities using syndromes. If we find the proportions of faults with identical adherence values, we can formulate a histogram as was done for detectability. The adherence histogram for the 74ls181 ALU is shown in Figure 9.4. These histograms were generally characterized by relatively low values of adherence except with sharp rises at the adherence value one. PO faults always have adherence values of one, but often an unexpectedly large proportion of other faults have this value as well.

9.2.2   Bridging Fault Behavior

As stated previously, Inductive Fault Analysis has shown that single stuck-at faults do not always model bridging faults very well [SHEN85]. This statement was made on the basis of faults that were injected into actual circuit layouts and then systematically filtered by an extraction process until they were mapped into various models (if possible). Our research has enabled us to make a similar statement, but from a completely functional standpoint, and independently of the circuit technology. Figure 9.5 shows the proportions of AND NFBFs and OR NFBFs which are equivalent to double stuck-at faults.

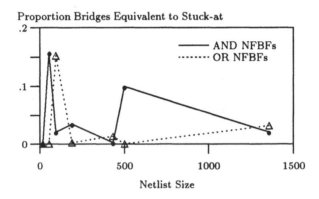

Figure 9.5: A plot illustrating the proportions of AND and OR NFBFs that exhibit stuck-at behavior.

The procedure for calculating these proportions is simple. The number of variables in the fault function (F_i, Definition 8.2) at the site of the bridging fault is counted. If that number is zero, then the bridging fault is either a logical zero or one, i. e. a stuck-at fault. For the circuits C499 and larger, functional decomposition was used to speed up Difference Propagation [HUNG89], so the fractions of NFBFs which are also double stuck-at faults for those circuits may not be completely accurate due to the decomposition masking some functional interactions. Our results agree with those of Inductive Fault Analysis in that these proportions of "stuck-at bridging faults" are generally low. Note also that circuits with relatively large proportions of AND NFBFs with stuck-at behavior have small numbers of OR NFBFs which are stuck-ats and vice versa.

In order to facilitate a fair comparison between bridging fault and stuck-at fault performance, plots analogous to Figures 9.1 through 9.3 are illustrated in Figures 9.6 through 9.8. In Figure 9.6, the detection probability histogram for the circuit 74ls181 was omitted as was its adherence histogram. In both cases the plots were difficult to read because of the large spread of values.

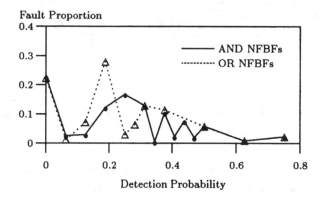

Figure 9.6: The bridging fault detection probability histograms for circuit C95.

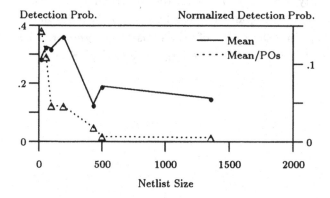

Figure 9.7: The trends of mean bridging fault detection probabilities and ratioed detection probabilities for the example circuits.

Figure 9.7 was not separated into bridging fault types because little difference was seen between AND and OR NFBFs. It can be seen that the mean

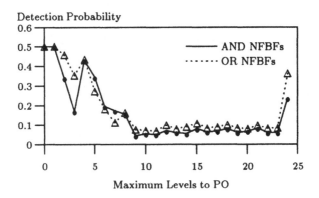

Figure 9.8: A plot of the mean bridging fault detectability versus maximum distance to POs for circuit C1355.

detectabilities of bridging faults are slightly higher than stuck-at means. It is also apparent that detectability is still decreasing with increasing circuit size.

In Figures 9.6 and 9.8, the behavior of AND and OR NFBFs are very nearly the same. In fact, most of the data gathered during our experimentation, with the exception of the stuck-at modelability study, indicate that the logic dominance value of the circuitry is of little consequence as far as the potential utility of the fault model or the detectability of the fault. Also, the NFBF adherence histograms differed little from the stuck-at adherence histograms except that the spread of values was usually greater. This is probably attributable at least in part to the larger sizes of NFBF sets.

Chapter 10

The Contributions of Controllability and Observability to Test

Structural testing is based on the notions of controllability and observability. Controllability generally refers to the application of circuit stimulus such that the presence of a fault will cause some site(s) in the circuit to be logically different than when the fault is absent. Observability concerns the causation of a condition whereby the difference at the site(s) would force a logical difference at some circuit output where it can be directly measured. These quantities will be more formally defined in a later section.

In this chapter we present exact measurements of controllability and observability for single stuck-at faults in a set of combinational circuits. The circuits used here are the same as those used in the fault model behavioral study of Chapter 9. We begin by formally stating the definitions of the quantities we measure. Next, we discuss the statistical methods used to measure the importance of controllability and observability in the fault detection process. Finally, we graphically illustrate the data and comment upon its implications. By correlating detectability with both controllability and observability we are investigating which component plays the more significant role in making circuits testable. The results are the first exact measures of these quantities for nontrivial examples. They suggest where design for test efforts should be concentrated and how to enhance test generation algorithms for better performance.

10.1 Motivation to Study Fault Controllability and Observability

Researchers have for many years been developing methods to study, and/or approximate the controllability and observability of faults in integrated circuits [PARK75], [GOLD79], [SAVI84], [MERC84], [FUJI90]. Due to the proliferation of the stuck-at fault model and the availability of scan design methodologies, the controllability and observability of single stuck-at faults in combinational circuits has been of particular interest. Unfortunately, even the assumptions of single faults and combinational circuits do not enable the approximate methods of testability measures to render a high degree of accuracy [AGRA82], [SAVI83], [MERC84]. Hence, test decisions based on these approximations may have questionable validity.

However, the successful application of design for testability and test strategies relies heavily on whether or not they attack the underlying problem which promises the larger incremental gain in individual fault detectability. This condition should be contrasted from the related question of which of controllability or observability is the easier condition to satisfy for the typical stuck-at fault. Most researchers would concede that observability is often more difficult of the two. But there is general disagreement about whether *enhancing* observability or controllability will lead to the most benefit.

By correlating detectability with both controllability and observability on an individual fault basis, we achieve a measure of the importance of each of the two latter parameters in encouraging fault detectability. Thus, we have a mechanism by which we can select the parameter upon which to concentrate our efforts in circuit modification for test purposes.

10.2   Definitions of Controllability and Observability

A fault in a combinational circuit is detected by an input vector iff the fault is simultaneously controllable and observable. Control involves the creation of a difference at the fault site in the fault free circuit value and the faulty circuit value (fault excitation). For the detection of the fault, the difference created at the fault site should be visible at an observable circuit output. Thus the detectability of the fault is intimately related to the ease with which the fault site can be controlled to a certain value and observed. In this section we define these parameters.

Definition 10.1 *The* detectability *of a fault is the fraction of all possible input vectors which detect the fault.*

Definition 10.2 *The* controllability *of a fault is the fraction of all possible input vectors which create a difference in the good circuit value and the faulty circuit value at the fault site.*

Definition 10.3 *The* observability *of a fault is the fraction of all possible input vectors which sensitize the fault site to at least one observable circuit output.*

In this chapter we study the relationships between these values for the single stuck-at fault model.

10.3 Generating Controllability and Observability Information with Difference Propagation

Difference Propagation has the capability of generating both controllability and observability information for logical faults. The details of how these calculations are accomplished are given in this section. Because these controllability and observability studies have only been carried out on stuck-at faults to date, we will discuss only the case of the stuck-at fault.

10.3.1 Calculating Fault Controllability Functions

As we saw in Section 8.1, functional calculations in Difference Propagation begin with a forward pass through the circuit to generate the good circuit function at each line as in CATAPULT. This function describes the value at that line for every possible input condition.

The controllability function of a fault is all those conditions in which a fault is excited. We represent the controllability function as a logic 1 for all input conditions which excite the fault and a logic 0 for all others. To excite a stuck-at-0 fault on a line, the line must be driven to a logic 1. The reverse is true for a stuck-at-1 fault. Thus, the good circuit function at a line also represents its controllability function for a stuck-at-0 fault, and the complement of the good circuit function is its controllability function for a stuck-at-1 fault. Difference Propagation recognizes this fact, and thus is implicitly able to find the controllability function (and its syndrome) for any stuck-at fault.

10.3.2 Calculating Fault Observability Functions

Calculating observability functions for stuck-at faults is slightly more complex than controllability functions. From Definition 10.3 in Section 10.2, the observability of a line is the fraction of minterms that cause the value of the line to directly influence the circuit outputs. Stated in terms of the concepts of Difference Propagation, it is the set of input conditions that cause a functional difference to propagate to all possible outputs. From Section 4.3.2, one way to calculate the observability function of a line would be to represent the line with its own variable in the control function calculations and then perform the Boolean difference. For large fault sets, this would require a great deal of costly composition and decomposition.

A method to achieve the equivalent results would be the following. Assume for a moment that we possess the capability of causing a stuck-at fault on some circuit line to be excited for all input conditions, i. e., the difference function at that line is a tautology (1). For most lines, there will only be certain input conditions such that the line will affect at least one circuit output. By using the propagation functions of Difference Propagation (Table 8.1), we can calculate the functions which describe the conditions when this completely excitable fault is observable. But, since there are no constraints for fault excitation, this is precisely the observability function of the fault.

This method can also be illustrated in terms of Figure 8.1. If we wish to calculate the observability function for stuck-at faults on line A in the single gate circuit, we set $\Delta f_A = 1$ and $\Delta f_B = 1$. Using Equation (8.9), we get

$$\Delta f_C = f_A \Delta f_B \oplus f_B \Delta f_A \oplus \Delta f_A \Delta f_B = f_B, \qquad (10.1)$$

which says that line A will be observable whenever line B is set to a logic one. This exactly describes the conditions which cause line A to be observable at the output C. If further gates exist in the path to the output, the process simply continues. Multiple output differences are simply ORed together just as in regular Difference Propagation.

10.4   Controllability/Observability Results and Analysis

Exhaustive data on the detectability, controllability and observability of the faults in our benchmark circuits has been generated using the algebraic methods discussed above. The faults used are single stuck-at checkpoint fault sets [BOSS71] minimized using fault equivalence [MCCL71]. Our methods for test set generation succeeded for 100% of the faults in all circuits except the C1908 from [BRGL85]. For this circuit, the technique succeeded for all but the 10% of the checkpoint faults closest to the primary inputs.

Figures 10.1 - 10.3 represent one measure of the interrelationships between these three parameters. Faults were first classed according to their detectability measures. The ranges of detectability used were 0-.05, .05-.1, .1-.15, and so on. The mean detectability, controllability, and observability values were calculated for each of these classes, and the results are plotted in Figures 10.1, 10.2, and 10.3 for the 74ls181 circuit, C432, and C1908, respectively.

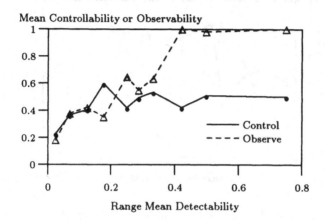

Figure 10.1: A plot of mean controllability and observability values versus mean range detectability for circuit 74ls181.

The lower ranges of detectability in Figures 10.2 and 10.3 indicate that low values of observability have a greater capacity to render a fault "hard to detect" than do low values of controllability. This is significant because faults which are hard to detect present the greatest obstacle to rapid test generation, fault simulation, etc.

To explain our last measure of correlation, refer to Figure 10.4. In this two dimensional plane we have defined two axes and four quadrants. The horizontal axis (representing the independent variable) corresponds to observability and the vertical axis to detectability (dependent variable). Assume that the two quadrants to the right of the vertical axis represent the event that the conditions for observation have been satisfied and to the left that they have not. Similarly, the quadrants above the horizontal axis represent the event that fault detection has occurred and below that no detection has occurred. We have therefore defined four mutually independent events represented by the four quadrants. Quadrant I represents a condition where both observation and detection have occurred. Quadrant II represents the impossible condition of a fault which is detected and not observed. Quadrants III and IV represent conditions which can occur, and are defined analogously to I and II.

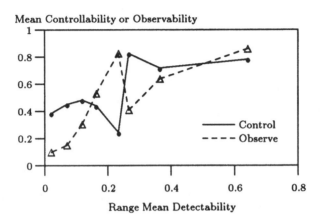

Figure 10.2: A plot of mean controllability and observability values versus mean range detectability for circuit C432.

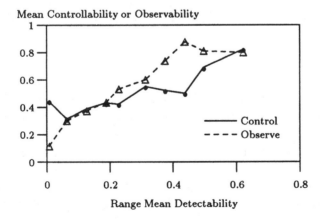

Figure 10.3: A plot of mean controllability and observability values versus mean range detectability for circuit C1908.

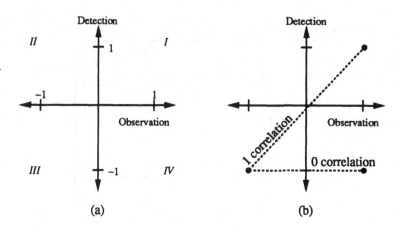

Figure 10.4: (a)The definition of detectability and observability quadrants. (b)Examples of perfect correlation and no correlation.

Now, for each test in the test set of a fault, we find the Quadrant to which it maps and increment the weight of "members" in that Quadrant (the weights are initialized to 0) at either (1,1), (1,-1), or (-1,-1). If, for example, upon exhausting the test set for our fault we find that only points (1,1) and (-1,-1) have nonzero weights, then the best fitting weighted least squares line would pass through these two points and would have a slope of one, as shown in Figure 10.4(b). That would indicate that for this particular fault, observability and detectability are perfectly correlated, i. e. observability implies detectability as well as the contrapositive. If, on the other hand, each of the points in the three possible quadrants has some nonzero weight, then the slope of the least squares line will indicate the degree to which observability causes detectability. If we continue this process for all faults in a particular circuit and then find the overall slope, then we have produced an aggregate measure of the same quantity.

The arguments stated above are just as applicable to the case of controllability and detectability. Figure 10.5 gives the values of the coefficients of correlation for both controllability and observability for each of the circuits in our benchmark set (represented by their respective netlist sizes). The resulting curves indicate more strongly than Figures 10.1 - 10.3 that observability is more correlated with detectability than is controllability for the larger circuits in our benchmark set.

Our interest in these test parameters is more on an individual fault ba-

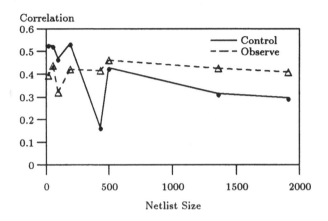

Figure 10.5: A quadrantized analysis of the correlations between controllability or observability and detectability versus netlist size.

sis. To study the correlations in finer detail we again categorized the faults according to their detectability. The plots for some of the benchmark circuits are shown in Figures 10.6 - 10.8.

These figures suggest that observability is more correlated with faults which are not very detectable and conversely that controllability has better correlation with highly detectable faults. In each case, the observability correlation of hard to detect faults is higher than that of controllability. Moreover, the detectability classes in which the observability correlation dominates is typically from 0.0 to 0.2 or 0.3. This makes sense because the least detectable faults are often near sites of high reconvergence where observability is always a problem and the most detectable faults are near the outputs where controllability is difficult [BUTL90b], [AGRA90]. Furthermore, the hard to detect faults frequently had good measures of both controllability and observability, which supports the thesis of Savir's study of these quantities [SAVI83].

10.5   Controllability/Observability Summary

We have sought to quantify the contributions of both controllability and observability to detectability for individual stuck-at faults as well as fault sets for traditional target stuck-at fault sets. The significance of the work is that it is the first of its kind to analyze detectability/controllability and detectability/observability both exactly and on a per-fault basis. Hence, arguments

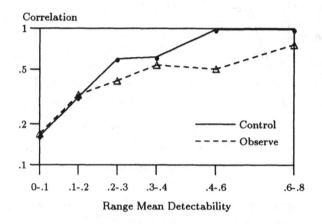

Figure 10.6: A quadrantized analysis of correlation between controllability or observability and detectability ranges for circuit 74ls181.

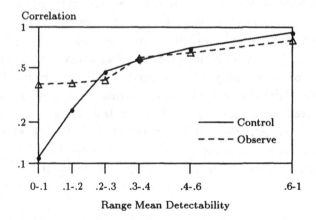

Figure 10.7: A quadrantized analysis of correlation between controllability or observability and detectability ranges for circuit C432.

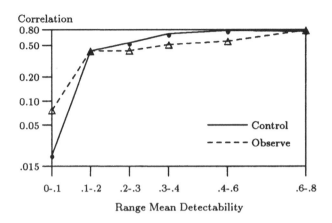

Figure 10.8: A quadrantized analysis of correlation between controllability or observability and detectability ranges for circuit C1908.

that the data are not correct due to approximations in the analysis do not apply. Furthermore, the faults studied cover the entire ranges of controllability, observability, and detectability, and the information about these regions is therefore complete.

The results have shown that observability is important for the most difficult to detect faults, which are always of concern in most design for test-related applications. Solutions to the design for test problem which enhance the the observabilities of these faults will thus have the most potential to simplify the test process. Furthermore, the statement that separate measures of either controllability or observability do not gauge testability is corroborated by our data [SAVI83]. Although our data concurs that there is no absolute relationship between the separate measures of controllability, observability and detectability, it suggests that faults can be categorized as "hard to detect" based on the point at which observability correlates better with detectability. In this research, that point appears to be relatively constant.

Chapter 11

Analyzing Test Performance with the ATPG Model

As we saw in the Introduction and again in Chapter 6, the quality of test sets is an important quantity, but one that is difficult to define and measure. Even if a suitable metric is defined, most existing methods to quantify the metrics fail to provide an acceptable level of statistical significance. In contrast, our method examines the properties of the entire set of one-vector-per-fault test sets. This number can be quite large.

For example, for the benchmark circuit C432, the mean number of vectors (of all 2^{36} input combinations) which detect a single stuck-at fault was found to be 3×10^9 [BUTL90b]. There are approximately 450 checkpoint stuck-at faults when they are minimized with gate equivalence. Assume that we employ fault simulation and fault dropping and that the final fault set whose test set is guaranteed to detect 100% of the single stuck-at faults consists of only 10% of the original 450 faults. In this case, the number of possible test sets would then be approximately $(3 \times 10^9)^{45}$.

To study each of these test sets stochastically, we use the combination of Difference Propagation and the ATPG model to measure fortuitous detection of non-target defects by complete target fault test sets. The circuits studied are the three smallest circuits from the combinational benchmarks [BRGL85], a fulladder circuit, and the 74LS181 ALU. The final analyses of these results suggest that stuck-at fault testing may not always be capable of producing acceptable quality as measured by defect level, which is defined in the next section.

This chapter is based on the paper "Quantifying Non-Target Defect Detection by Target Fault Test Sets", which first appeared in the *Proceedings of the 2nd European Test Conference*, Munich, Germany, April 10-12, 1991, pp. 91-100.

11.1   Defect Level Motivation

In Chapter 5, defect level was defined as the proportion of circuits passing all phases of manufacturing test which are actually faulty. For large circuits, defect levels today are typically on the order of 200 DPM. However, the Semiconductor Research Corporation (SRC) has set forth the ambitious goal of 1 DPM in circuits of 10^7 devices [SRC85]. This goal requires extremely high performance methods of test as well as good measures of test quality.

A defective circuit will pass the manufacturing test if the following two events both occur for at least one irredundant defect.

1. The defect is present in the circuit.

2. The test set generated for the circuit does not detect the defect when it is present in the circuit.

These two events can be assumed to be independent. Let F be the event that a circuit is defective. Let p_i denote the probability that defect i is present in a circuit (corresponding to Item 1). Let m_i denote the probability that a test set generated for the circuit fails to detect defect i given that it has occurred (corresponding to Item 2). Due to the assumption of independence, it follows that the probability that a circuit is defective due to defect i is

$$P(F) = p_i m_i. \tag{11.1}$$

If the number of single defects that can occur is M, then the probability of a defective part due to a single defect is then

$$P(F) = \sum_{i=1}^{M} p_i m_i. \tag{11.2}$$

If all the p_i are very small ($<< 1$) and defects occur at sites independently of one another, then the probability that a part is defective can be approximated by the probability that a part is defective due to a single defect, i. e., $P(F) \approx P(F)_1$. When multiplied by 1×10^6, $P(F)$ is actually the defect level expressed in DPM. Note that the sum in Equation (11.2) can be split into two parts, the contribution due to target faults and the contribution due to non-target defects. Let M_T be the number of target faults and M_N the number of non-target defects. Then $M = M_T + M_N$ and

$$P(F) = \sum_{j=1}^{M_T} p_j m_j + \sum_{k=1}^{M_N} p_k m_k \tag{11.3}$$

If a test set will be generated which covers all the faults in the target set, then the leftmost sum on the right side of Equation (11.3) is identically 0 because each of their respective $m_j = 0$. The question that remains to be answered is that of the contribution of non-target defects to defect level. Certainly some non-target defects will be detected by the target fault test sets, but just how likely are these fortunate occurrences?

11.2   ATPG Model Development

To further illustrate the idea of fortuitous non-target defect detection, consider the example of Figure 11.1. Assume that this figure represents a simple system in which only two defects could ever occur. The region labeled T represents the complete test set for the defect that matches our target fault for test generation. The region labeled N is the test set for the other defect that could occur, but is not a target for test generation due to economic concerns or because our ATPG system cannot generate tests for this type of defect. The shaded region $(N \cap T)$ represents the portion of tests for the target fault which will fortuitously detect the non-target defect as well. If $N \cap T$ is very small, it is highly likely that the test set we generate for this circuit via ATPG, be it random, deterministic, or a hybrid, will not also detect occurrences of defect N. If the non-target defect occurs with any frequency, then such a test would result in defective circuits which pass the test.

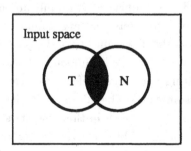

Figure 11.1: A Venn diagram illustrating the relationship between target and non-target defect test sets.

More precisely, define A as the event that the non-target defect is fortuitously covered by the target fault test set. We will employ the following model of a single test vector generation event.

Definition 11.1 *An ATPG event consists of the random selection of a single vector from the entire test set for the target fault.*

Using our ATPG event model, we can calculate the probability of $\overline{A}$, the event that the vector we generate or "select" for our target fault will not detect or will "miss" the non-target defect. Graphically, this is simply the ratio of the unshaded portion of T to the whole region T in Figure 11.1, or

$$P(\overline{A}) = \frac{|\overline{N} \cap T|}{|T|}. \qquad (11.4)$$

where $|S|$ is the cardinality of test set S. To complete the ATPG model, we must extend our event model to the general case of multiple target fault sets.

Definition 11.2 *The ATPG process consists of independent occurrences of ATPG events.*

Our ATPG model is arguably an approximation to actual ATPG systems. Its aim is not to capture the exact behavior of true ATPG algorithms, but rather to emulate the essence of ATPG in a computationally tractable fashion. Our assumption of the independence of ATPG events amounts to assuming that no fault simulation occurs between test generation events. Modeling the use of fault simulation requires fault ordering which adds system specificity and a great deal of complexity to the problem. More importantly, fault simulation is normally used in conjunction with fault dropping, thus shortening target fault sets. Because test sets for target fault sets behave essentially randomly with respect to the detection of non-target defects, the practice of fault dropping actually decreases the likelihood of detecting them. Our desire for optimism in estimating non-target defect coverage leads us to eliminate the fault simulation assumption. This allows us to be optimistic about the quality of the target fault sets (which in this case are single stuck-at faults). If these test sets show unacceptable behavior under our assumptions, then it is very optimistic to expect better performance from 100% coverage single stuck-at test sets in practice.

Deterministic ATPG algorithms typically generate partially specified vectors (X values in the vector) rather than completely specified vectors (with only 0's and 1's in the vector) as in our model. However, these partially specified vectors are often "filled out" to increase coverage of other target faults (by replacing the X's with 0's and 1's). Randomly generated patterns (the majority of ATPG patterns) are almost always completely specified. Also, our focus in this research is on the least detectable defects, whose test sets are very small. These highly untestable defects are more likely to be hard to detect with tests generated for the target fault set. In both these situations, our single vector assumption is consistent with our goal of optimism in estimating test quality. Finally, most ATPG is not truly random, but is guided by heuristics. Because these heuristics are based upon target fault model assumptions, the additional constraints imposed by non-target defects are satisfied fortuitously, if at all. These events can best be approximated as occurring randomly.

This model constitutes a figure of merit on the utility of the test set. The higher the likelihood that non-target defects occur, the more meaningful is the measure of test quality. The Inductive Fault Analysis research has indicated that these likelihoods seem to be increasing for certain classes of defects [SHEN85], [FERG88]. More importantly, the lower the acceptable defect level, the more critical it becomes to account for non-target defects, even if their occurrences are infrequent.

11.3 Fault Set Selectability

The definition of the ATPG model permits more detailed study of how test sets for individual faults interact with one another. The mechanism of this interaction is the overlap of minterms in the test sets. Since some minterms will cover more faults than others, those minterms will likely be more "selectable" than the others. We will thus associate the term *selectability* with the fault model's causation of some minterms to become preferential over others.

To more fully explain the idea of fault selectability, consider the example of Figure 11.2. The combinational circuit in part (a) of the figure contains the single stuck-at fault A stuck-at-1. With the application of some Boolean algebra, one would learn that three vectors exist which are valid tests for this logical fault, $\overline{AB\overline{C}\,\overline{D}}$, $\overline{AB\overline{C}}D$, and $\overline{ABC\overline{D}}$. Under the ATPG event model, each of these vectors are equally likely to be selected as a test for the fault, so each of the vectors has a selectability of $\frac{1}{3}$. This is illustrated in Figure 11.2(b) by placing the fraction $\frac{1}{3}$ in the corresponding minterms in the Karnaugh map of the function realized by the circuit in part (a) of the figure.

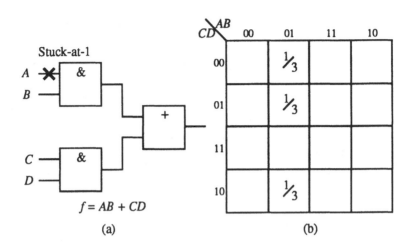

Figure 11.2: (a) An example circuit containing a single stuck-at fault. (b) The fault selectability under the ATPG model.

If M_i is the event that minterm i is selected as a test for the single fault and S is the set of tests for the fault, then the selectability of minterm i is

$$P(M_i) = \frac{1}{|S|} \ \forall \ i \in S. \tag{11.5}$$

Since A stuck-at-1 is the only fault in this example and none of the other minterms are tests for this fault, the remaining 13 minterms have selectabilities of 0. This is expressed by blank entries in the corresponding minterms in the Karnaugh map.

Moving on to a slightly more complex example, consider the case of two faults in the fault set. In Figure 11.3(a), our example circuit now shows the fault set containing two single stuck-at faults, A stuck-at-1 and C stuck-at-1. Each of these faults has a test set with three minterms. The analysis is very similar to the one of our previous example. The non-overlapping terms have selectabilities of $\frac{1}{3}$ as can be seen from the Karnaugh map of Figure 11.3(b). The only difference from the single fault case is that now the two test sets overlap in the minterm $\overline{A}B\overline{C}D$. Fortunately, our ATPG process model can take care of such occurrences. The selectability of the minterm at which overlap occurs is the probability that it will be generated as a test for either of the faults. This is equivalent to 1 minus the probability that it will never be selected. Each of these "not select" probabilities is $\frac{2}{3}$ by our ATPG event model.

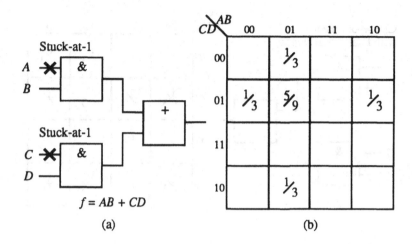

Figure 11.3: (a) An example circuit containing two single stuck-at faults. (b) The fault set selectability under the ATPG model.

Let L be the number of faults in the fault set and S_j be the test set for the fault j. Because vector generations are independent by the ATPG process model, the selectability of minterm i is

$$P(M_i) = 1 - \prod_{j=1}^{L} \left(\frac{|S_j| - |S_j \cap M_i|}{|S_j|} \right).$$ (11.6)

In our example, the selectability of minterm $\overline{A}B\overline{C}D$ is $1 - \left(\frac{2}{3}\right)^2 = \frac{5}{9}$.

The notion of fault selectability can theoretically be generalized to any set of faults, circuit type, and ATPG environment. It also has interesting implications in terms of definition of the problem of test performance evaluation of fault sets. The conclusion derived from fault selectability can be stated as

Statement 11.1 *Every fault model infers a distribution of probabilities over the input space of the function.*

The selectability distribution of probabilities is given for our example circuit in Figure 11.4(b). The fault set in this case is the set of checkpoint faults for the circuit which has been minimized using fault equivalence.

The realization of distributions of probabilities brings us one step closer to the solution of test performance evaluation and perhaps enhancement of test performance. Test performance can be calculated by using the distribution

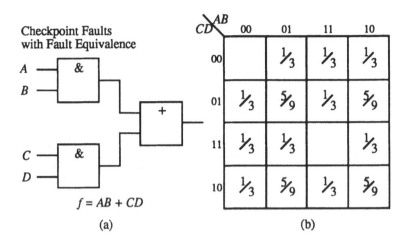

Figure 11.4: (a) An example circuit containing the set of checkpoint single stuck-at faults minimized using fault equivalence. (b) The fault set selectability distribution of probabilities.

of probabilities to discern how likely some defects are of being tested or not tested. Test performance can be enhanced by maximizing these probabilities for that subset of defects deemed critical. Certainly, depending on the complexity of the ATPG system or model being employed, the probabilities can be calculated with varying degrees of difficulty. Our ATPG model permits the selectability distribution of probabilities for logical faults to be calculated relatively easily.

11.4 Probabilistic Non-Target Defect Coverage

The ATPG model defined in earlier sections provides a framework for study of fortuitous detection probabilities of non-target defects. The process of calculating probabilistic coverage values is illustrated in the example below.

Assume that we wish to generate a test set for the simple circuit shown in Figure 11.5(a), which is repeated from Figure 11.4(a). In order to use ATPG we will select a target fault set and provide a circuit description and the fault set to our ATPG system for test set generation. Checkpoint stuck-at faults with gate equivalence as defined previously provide 100% coverage of all single stuck-at faults in the circuit, so we select this fault set as our target set. For our example, the target fault set is then $\{A_0, A_1, B_1, C_0, C_1, D_1\}$, where a fault I stuck-at-x is denoted I_x. The faults for which each input combination is a test are listed in the Karnaugh map entry for that input combination in Figure 11.5(b).

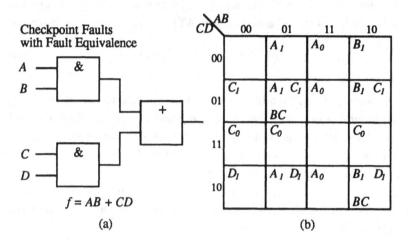

Figure 11.5: (a) An example circuit for ATPG. (b) The corresponding target and non-target fault test sets.

Now assume that although we cannot afford further test generation effort, we are concerned about the prospect of a wired-OR type bridging fault between inputs B and C. The input combinations which would detect the non-target defect are indicated analogously in Figure 11.5(b) with the designation BC. Each stuck-at fault whose test set "overlaps" with the test set for the bridging fault has the potential of having the vector generated for it fortuitously detect the bridging fault. Thus, by finding the cardinalities of the stuck-at fault test sets and the intersections of each of these stuck-at fault test sets with the bridging fault test set, we can calculate the probability that each stuck-at fault's test vector detects the bridging fault by applying a result derived from Equation (11.6). We begin by defining the set of four events, I, $\forall\ I \in \{A, B, C, D\}$ and the event T.

$I \triangleq$ Event that a test generated for fault I_1 also
detects the bridging fault BC.

$T \triangleq$ Event that the test set generated for the target
faults also detects the non-target defect BC.

We only need to define events for the stuck-at-1 faults because in our example only the stuck-at-1 fault test sets overlap with the bridging fault test set. As in many probability calculations, it is actually simpler to calculate the probability of the event $\overline{T}$, the event that the target fault test set fails to detect occurrences of non-target defect BC, which we call the *miss probability*. The ATPG process model assumes that ATPG events are independent, so

$$P(\overline{T}) = \prod_{I \in \{A,B,C,D\}} P(\overline{I}). \tag{11.7}$$

In other words, the probability that the test set fails to detect the bridging fault is the probability that the vector generated for each overlapping stuck-at fault does not detect the bridging fault. Because vector generations are assumed independent, this is simply the product of the individual probabilities. Now, we need to calculate each of these individual probabilities of missing the non-target defect. These calculations can be performed in much the same way that we calculated probabilities of not selecting a particular minterm in Equation (11.6).

ATPG events are modeled as random test vector selections. By the Karnaugh map of Figure 11.5(b), the test set for fault A_1 has three members, only one of which is also a test for the bridging fault BC. Thus, there are two chances in three that we will generate a vector for the fault A_1 which does not detect the bridging fault BC, so the probability of that event is $\frac{2}{3}$. The same is true of the other three overlapping target faults by symmetry so

$$P(\overline{I}) = \frac{2}{3} \ \forall \ I \in \{A, B, C, D\}. \tag{11.8}$$

Substituting these values into Equation (11.7) we have

$$P(\overline{T}) = (\frac{2}{3})^4 = 0.1975, \text{and} \tag{11.9}$$

$$P(T) = 1 - P(\overline{T}) = 0.8025. \tag{11.10}$$

If we define C_k as the event that a non-target defect k is covered by a target fault test set whose cardinality is L, then the procedure for the calculation of non-target defect coverage probability can be formally stated as

$$P(C_k) = 1 - \prod_{j=1}^{L} \left(\frac{|S_j| - |S_j \cap S_k|}{|S_j|} \right). \tag{11.11}$$

The complement of C_k for a non-target defect is in fact its miss probability, or the probability that the target fault test set will not detect it if it occurs.

The result from our example indicates that there is an 80% chance that the test set we generate for the stuck-at faults will cover the bridging fault as well. We have thus produced a measure of the target fault test set's coverage effectiveness for this particular non-target defect. The measure calculated by complementing the result of Equation (11.11) is actually an optimistic estimate of the miss probability of the non-target defect by any test set for the target faults which consists of no more than one test vector per target fault. We are then in some sense measuring the inherent "elusiveness" of non-target defects over all target fault test sets.

11.5 Faults Sets

Having developed the ATPG model and the complete test set information, it is necessary to select sets of target faults and non-target defects for study. The target faults should obviously be stuck-at faults, given their wide acceptance in industry. The particular stuck-at fault sets chosen here are checkpoint faults which were minimized using simple gate equivalence rules [BOSS71], [MCCL71]. Checkpoint faults were selected due to their frequent use as a target fault set. Gate equivalence was applied to minimize duplication of test set computations.

The non-target defect sets used here are bridging faults. Bridging faults at this time seem to be the second most dominant fault model studied in the literature, single stuck-at faults being the first. This is largely because empirical research has indicated that many physical failure mechanisms manifest themselves as bridges [GALI80], [SHEN85]. The particular instances of bridging faults we have selected are the two line, non-feedback variety. Two-line faults are commonly used with the assumption that bridging faults of more than two lines are highly unlikely. We also desire our results to be as compatible with the literature as possible. NFBFs were chosen primarily because empiricism has indicated that FBFs are much more likely to be detected by stuck-at fault test sets than are NFBFs [MEI74], [MILL88]. Just as in the fault behavior study (Chapter 9), this work includes both AND NFBFs and OR NFBFs to study the effect of the dominant logic value on test performance.

The NFBF sets were reduced using the results of Theorem 2 of Mei [MEI74] to remove all NFBFs between two inputs to the same gate. Mei's theorem proves that test sets with 100% single stuck-at fault coverage, such as checkpoints fault test sets, also cover all NFBFs between inputs to the same gate. All statistics reported in this work are for these reduced sets of NFBFs.

11.6   Test Performance Results

The analysis detailed in the previous section has been carried out on the set of combinational benchmark circuits listed at the end of the introduction to this chapter. Difference Propagation is used to generate complete test sets for both the target faults and non-target defects. Because individual fault test sets are represented symbolically as OBDDs, target fault/non-target defect test set intersections can be found by simplying ANDing the functions describing the target fault and non-target defect test sets using Bryant's *Apply* algorithm [BRYA86]. We can then find all the cardinalities used in Equation (11.4) by a simple minterm counting algorithm similar in nature to the one presented by Bryant.

As stated previously, Mei's Theorem 2 was used to eliminate all NFBFs theoretically guaranteed to dominate stuck-at faults [MEI74]. Figure 11.6 graphically shows the proportions of the remaining NFBFs (both AND and OR) which domi⁻.ate at least one fault in the target fault set.

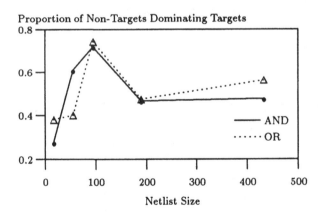

Figure 11.6: A plot of the proportions of non-target defects dominating target faults for five benchmark circuits.

This measure is significant because these proportions of non-target defects are guaranteed to be detected by test sets which cover the corresponding target faults. No clear trend for non-target defect fault domination seems apparent, but significant proportions of NFBFs are shown not to be guaranteed covered by 100% coverage single stuck-at fault test sets for each of the circuits in this study.

Figures 11.7 and 11.8 show the mean and maximum miss probabilities of non-target defects calculated under the ATPG model. It is interesting to note that no NFBFs were found which would definitely escape detection should they occur. For circuits larger than the 74LS181, a fault sampling procedure had to be introduced due to the enormous number of candidate NFBFs in these circuits. This sampling process is described in [BUTL90b]. Thus, as circuit size increases, we are sampling increasingly smaller proportions of the entire population of non-target defects. Such a phenomenon obviously increases the likelihood that we will not locate the "most easily missed" non-target defect(s). However, our sample data still reveals non-target defects which exhibit relatively high probabilities of being missed. This is suggestive that there may be cause for concern about test quality in the not-too-distant future.

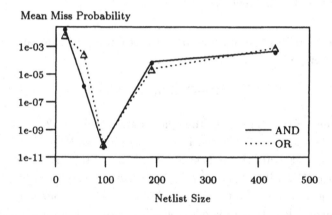

Figure 11.7: A plot of mean miss probabilities versus netlist size under the ATPG model for the benchmark circuits.

11.7 Implications to Defect Level

Using the notation of Section 11.1, we have succeeded in estimating values of m_i for non-target defects. If we can now find values for p_i, we can compute DL by using Equation (11.3). In Williams' and Brown's famous derivation of the defect level equation, they assume the p_i has a uniform and equiprobable distribution [WILL81]. So p_i is a constant, p. The yield, Y, of integrated circuits is the proportion of ICs produced which are fault free. For a circuit

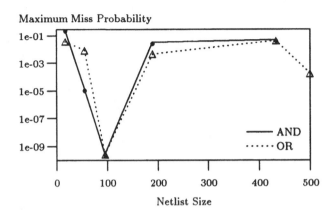

Figure 11.8: A plot of maximum miss probabilities versus netlist size under the ATPG model for the benchmark circuits.

with n potential fault sites and probability p of faults occurring at each site, Williams and Brown give the yield as

$$Y = (1 - p)^n. \tag{11.12}$$

Again, assume a circuit similar to the benchmark circuit C432. If we assume that $Y = 0.95$ and $n = 450$, Equation (11.12) gives $p=1.14 \times 10^{-4}$. For the single NFBF in C432 with the highest miss probability, this gives a defect level of just under 6 DPM. If the largest miss probability is very much larger than the remaining miss probabilities, as has been the case in our experiments, then this single calculation provides a good approximation of the expected defect level under conditions similar to these. This defect level would be quite acceptable, but depending on the relations of m_i and p to circuit size, the problems may be growing worse. In particular, 6 DPM is somewhat large for a circuit the size of C432 and one could only expect that larger circuits would exhibit higher defect levels. It should also be noted that other classes of defects may exist which are even harder to detect with stuck-at test sets than bridging faults.

Chapter 12

Conclusions

The rapid advance of integrated circuit technology has made possible the fabrication of devices containing literally millions of devices. The associated problem of testing such complicated circuits and correctly certifying as "good" only those which are truly fault free has thus been compounded. In this environment, testing costs are now a significant proportion of the costs of fabricating ICs. The economics of this situation places a high priority on obtaining the highest performance possible out of automated test methods.

Running counter to the demands placed on test performance are the facts that

1. The aging stuck-at fault model may not be well matched to common defects in currently popular IC technologies, and

2. There currently exists no accurate methods for measuring test set performance that produce reliable, statistically significant results.

In this research we have introduced a method of test performance evaluation which addresses both of these problems. Most prior research has investigated Problem 1 by examining the frequency with which fault model isomorphisms occur. We propose that a second fundamental question must be probed relating to the fault model's ability to cause ATPG to derive test sets which fortuitously detect instances of non-target defects. Non-target defects are those defects which might occur but are not part of the target fault set given to the test generation system. Because defects which cause system failures will often manifest themselves as something other than the target fault type, these "other" classes of defects are very important.

Our solution to the second problem above consists of probabilistic measure of defect coverage for large numbers of test sets. In order to study all these different potential test sets stochastically, we must map out the "topology" of

the individual fault test sets. That is, we must calculate not only the sizes of
the test sets for individual faults, but also their propensity to detect whatever
non-target defects are of concern.

This need has inspired the development of a Boolean functional test generation system called Difference Propagation. Difference Propagation is not
proposed as a replacement to current ATPG methods for combinational circuits. Rather, it is an attractive solution to the test *set* generation problem
in an ambitious approach such as this one.

Difference Propagation has several distinct characteristics. First of all,
the functions represented are the difference between the good and defective
machines. Empiricism indicates that this is the preferred approach from a
computational point of view. Because the Boolean mathematics underlying
the development of Difference Propagation do not include assumptions about
the nature or multiplicity of the fault model other than a logical manifestation,
it can address more than just the single stuck-at fault model. Furthermore,
the method does not require the expensive functional decomposition and composition that is necessary in other methods employing the Boolean difference.
However, limited decomposition can and has been incorporated effectively to
speed functional calculations, as was shown in the minimum width decomposition work detailed in Section 8.5.3.

Prior to the use of Difference Propagation in its intended application, it
has been utilized to perform several interesting studies. Fault model behavior
was scrutinized using the method. The results and analysis revealed some
interesting properties of both classical (stuck-at) and non-classical (bridging)
fault in digital circuits. Some results were intuitive, and some not-so-intuitive.
To the knowledge of the authors, none of the results given in this phase of
the research have ever been investigated with exact methods except for trivial
examples.

With some slight modifications, Difference Propagation was also made to
produce results quantifying the controllability and observability of faults. The
first such investigations have concerned stuck-at faults. Again, the data illumi-
nated interesting fault parameters. Fault detectability was shown empirically
to correlate with observability for low detectability values and controllability for high detectability values. Furthermore, the overall correlation of detectability with observability seemed to be consistently better than with controllability. This information indicates that design for testability approaches

should focus on enhanced observability. Similar studies revealing quite different results have since been carried out for bridging faults [KAPU91].

The implementation of Difference Propagation has relied on the concept of Ordered Binary Decision Diagrams. While many methods for functional representation exist, the OBDD has shown the most promise when performing automatic manipulation of arbitrary functions. OBDDs have been shown to have unacceptable computational complexity for certain functions, but those cases have been conjectured by most authors to be rare. In trying to quantify a response to the query of just how good are OBDDs, the authors along with other researchers at The University of Texas selected a test case of a class of functions, symmetric functions, which lend themselves to very efficient representation using OBDDs. An unprecedented set of equations were derived which give the exact size of the representation for subclasses of symmetric functions as functions of integers which completely specify the particular symmetric functions. Because the representations become quite large for even modest symmetric functions, the results indicate that the straightforward application of OBDDs to large functional representation problems may often be unsuccessful.

The functional methods outlined in the previous chapter are ideally suited to an investigation of test performance, but they cannot achieve this goal alone. The largest drawback to currently proposed methods of test performance evaluation is that their results may be questionable due to lack of statistical significance. Unfortunately, the number of test sets in typical applications is so vast that it would be difficult to enumerate enough samples to obtain an acceptable degree of significance. In this work, a probabilistic model of ATPG has been proposed to resolve this issue. The model is not an exact analytical model of all ATPG systems, but rather one that mimics the essential components of ATPG. This model combined with Difference Propagation enables one to explore the quality of test sets through the probabilistic examination of a large number of them. It is expected, then, that the estimates of "miss probability" of certain non-target defects will be accurate estimates. The results of this analysis have provided evidence that conventional stuck-at testing may be approaching the defect level limit beyond which the quality levels arising from its use will not be acceptable to IC manufacturers or consumers.

Chapter 13

Suggestions for Future Research

In the course of these investigations, many interesting questions have arisen that, due to time considerations, will remain unanswered in this research. This chapter will explore some of these questions in more detail.

13.1 Extensions to OBDD Size Research

Credence could be added to the OBDD size studies if it could be shown that symmetric functions do indeed produce some of the most compact OBDDs amongst all functions of the same inputs. Furthermore, size equations for other classes of functions would also be quite helpful in deciding the fate of the representation. Equations giving OBDD sizes for integer multipliers of certain widths were recently published [BURC91b]; more would be useful.

13.2 Extensions to Difference Propagation

Difference Propagation would be made more attractive if further enhancements were made to improve its time and space performance. Recent publications have indicated that OBDDs can likely be manipulated more efficiently than the system currently in use [BRAC90]. Moreover, Difference Propagation might be applied to other instances of fault models for behavioral and controllability and observability analyses. With some extensions, these fault models might include faults which are not unidimensional, such as delay faults or sequential faults. The decomposition employed in Difference Propagation has extended its usefulness, but to date remains somewhat nebulous. Further studies to rigorously define exactly those conditions where and how many decompositions should be performed would be an aid to all such functional methods, as suggested in Section 8.5.4. The conjecture here is that the answer is related to the amount of reconvergence that occurs at the candidate minimum width points.

13.3   Extensions to Test Quality Research

Several additional studies could be useful to further evaluate test performance. One would be to turn the tables and calculate miss probabilities of stuck-at faults by other potential target defect sets. Depending on the defect set chosen, this would require the identification of a set of defects analogous to checkpoint stuck-at faults which have some acceptable coverage properties of all or most defects of their type. The addition of the fault simulation assumption to the ATPG model might produce more realistic estimates of defect miss probabilities. This addition would only be acceptable if it could be done in such a way as to preserve independence of the method from specific ATPG systems, i. e., resolving the fault ordering problem. Finally, to obtain results more indicative of the parameters of a particular design and fabrication environment, one could select faults based on the extraction of defects from actual layout information.

13.4   Using Ordered Partial Decision Diagrams

Yet another extension which might dramatically change the scope of applicable circuit size for much of the research detailed in this monograph would be the assimilation of Ordered Partial Decision Diagrams (OPDDs) [ROSS90], [ROSS91b]. OPDDs are a generalization on the concept of OBDDs. They permit the representation of partial functional information through the use of controlled breadth first expansion (as opposed to the depth first expansion generally used in OBDDs) and a third terminus, the *unknown terminus*.

The unknown terminus is similar to the 0 and 1 termini of an OBDD. Paths which end in the unknown terminus represent sets of minterms whose functional result has not been determined, i. e. each minterm is not known to cause the function to evaluate as either a 0 or a 1, hence the name "unknown". OPDDs which contain at least one path terminating in the unknown terminus represent *partially specified* functions. When no unknown terminus is present in an OPDD, it is a *fully specified* representation of a function and is, in fact, an OBDD as well.

OPDDs are created and operated upon in much the same way that OBDDs are. First, one must predetermine a variable ordering. Next, an OPDD size limit, called k_{apply}, must be selected based on a rough estimate of the memory resources than can be consumed during functional calculations. Then,

OPDDs are generated in the standard way using an *Apply* algorithm which expands breadth first as mentioned previously.

If the OPDD resulting from applying a Boolean operator to two fully specified operand OPDDs stays within the imposed size limit during functional expansion, then a fully specified OPDD is created. If on the other hand k_{apply} is exceeded, then pruning of some of termini is required. A greedy heuristic is used which seeks to keep only those vertices which are in the shortest paths between the root and the 0 and 1 termini. This criterion is used because those paths contain the most amount of functional information in the least amount of space [ROSS90].

The implementation of an OPDD calculation system has demonstrated that dramatic improvements can be made in the throughput, the number of minterms per CPU second, of functional calculations using OPDDs rather than OBDDs. [ROSS90]. For many applications, the partial results obtained from OPDDs is enough to do useful work [ROSS91b]. For other applications where fully specified information is required, such as design verification, one may iterate on the remaining unspecified information.

Clearly OPDDs could be used to replace OBDDs as the engine for the functional calculations carried out in this research. An interesting example would be to use OPDDs to calculate test quality. Whereas before the miss probabilities were calculated exactly, the use of OPDDs would result in miss probability intervals [BUTL91b]. Depending on the computer resources available, the value of k_{apply} and the number of iterations could both be increased so as to tighten the interval boundaries and hopefully converge on exact probabilities. Other extensions using OPDDs would be similar.

13.5 General Extensions

All the analyses detailed herein would benefit from being applied to further benchmark cases. Furthermore, unless full scan methods become completed integrated into all IC design environments, the methods might gain from an extension to the sequential circuit domain.

Bibliography

[ABAD86] M. S. Abadir and H. K. Reghbati, "Functional test generation for digital circuits described using binary decision diagrams," *IEEE Trans. Comput.*, vol. C-35, no. 4, pp. 375–379, Apr. 1986.

[ABRA80] M. Abramovici and M. A. Breuer, "Multiple fault diagnosis in combinational circuits by single fault test sets," *IEEE Trans. Comput.*, vol. C-29, no. 6, pp. 451–460, June 1980.

[ABRA83] M. Abramovici and P. R. Menon, "A practical approach to fault simulation and test generation for bridging faults," in *Proc. 1983 IEEE Int. Test Conf.*, Oct. 1983, pp. 138–142.

[ABRA86] M. Abramovici, P. R. Menon and D. T. Miller, "Checkpoint faults are not sufficient target faults for test generation," *IEEE Trans. Comput.*, vol. C-35, no. 8, pp. 769–771, Aug. 1986.

[ACKE83] J. M. Acken, "Testing for bridging faults (shorts) in CMOS circuits," in *Proc. ACM/IEEE 20th Design Automation Conf.*, June 1983, pp. 717–718.

[AGAR81] V. K. Agarwal and A. S. F. Fung, "Multiple fault testing of large circuits by single fault test sets," *IEEE Trans. Comput.*, vol. C-30, no. 11, pp. 855–865, Nov. 1981.

[AGRA75] P. Agrawal and V. D. Agrawal, "Probabilistic analysis of random test generation method for irrendundant combinational logic networks," *IEEE Trans. Comput.*, vol. C-24, no. 7, pp. 691–695, July 1975.

[AGRA82] V. D. Agrawal and M. R. Mercer, "Testability measures - What do they tell us?," in *Proc. 1982 IEEE Int. Test Conf.*, Nov. 1982, pp. 391–396.

[AGRA85] V. D. Agrawal, S. C. Seth and C. C. Chuang, "Probabilistically guided test generation," in *Proc. IEEE Int. Symp. on Circ. Syst. (ISCAS)*, June 1985, pp. 687–690.

[AGRA90] V. D. Agrawal, personal communication, June 1990.

[AKER59] S. B. Akers, Jr., "On a theory of Boolean functions," *J. Soc. Indust. Appl. Math.*, vol. 7, no. 4, pp. 487–498, Dec. 1959.

[AKER77] S. B. Akers, Jr., "On the specification and analysis of large digital functions," in *Dig. Papers, 7th Int. Symp. Fault-Tolerant Comp.*, June 1977, pp. 88–93.

[AKER78a] S. B. Akers, Jr., "Functional testing with binary decision diagrams," in *Dig. Papers, 8th Int. Symp. Fault-Tolerant Comp.*, June 1978, pp. 75–82.

[AKER78b] S. B. Akers, Jr., "Binary decision diagrams," *IEEE Trans. Comput.*, vol. C-27, no. 6, pp. 509–516, June 1978.

[BANE84] P. Banerjee and J. A. Abraham, "Characterization and testing of physical failures in MOS logic circuits," *IEEE Design and Test of Comput.*, vol. 1, no. 4, pp. 76–86, Aug. 1984.

[BEH82] C. C. Beh, K. H. Arya, C. E. Radke and K. E. Torku, "Do stuck fault models reflect manufacturing defects?," in *Proc. 1982 IEEE Int. Test Conf.*, Nov. 1982, pp. 35–42.

[BERM88] C. L. Berman, "Circuit width, register allocation, and reduced function graphs," IBM Res. Rep. RC 14127, Nov. 1988.

[BHAT84] B. B. Bhattacharya and B. Gupta, "Logical modeling of physical failures and their inherent syndrome testability in MOS LSI/VLSI networks," in *Proc. 1984 IEEE Int. Test Conf.*, Oct. 1984, pp. 847–855.

[BOSS71] D. C. Bossen and S. J. Hong, "Cause-effect analysis for multiple fault detection in combinational networks," *IEEE Trans. Comput.*, vol. C-20, no. 11, pp. 1252–1257, Nov. 1971.

[BRAC90] K. S. Brace, R. E. Bryant and R. L. Rudell, "Efficient implementation of a BDD package," in *Proc. ACM/IEEE 27th Design Automation Conf.*, June 1990, pp. 40–45.

[BRAH84] D. Brahme and J. A. Abraham, "Functional testing of microprocessors," *IEEE Trans. Comput.*, vol. C-33, no. 6, pp. 475–485, June 1984.

[BRGL85] F. Brglez and H. Fujiwara, "A neutral netlist of 10 combinational benchmark circuits and a target translator in FORTRAN," in *Proc. IEEE Int. Symp. on Circ. Syst. (ISCAS)*, June 1985, pp. 695–698.

[BRYA86] R. E. Bryant, "Graph-based algorithms for Boolean function manipulation," *IEEE Trans. Comput.*, vol. C-35, no. 8, pp. 677–692, Aug. 1986.

[BRYA87] R. E. Bryant, D. Beatty, K. Brace, K. Cho and T. Sheffler, "COSMOS: A compiled simulator for MOS circuits," in *Proc. ACM/IEEE 24th Design Automation Conf.*, June 1987, pp. 9–16.

[BRYA91] R. E. Bryant, "On the complexity of VLSI implementations and graph representations of Boolean functions with application to integer multiplication," *IEEE Trans. Comput.*, vol. C-40, no. 2, pp. 205–213, Feb. 1991.

[BURC91a] J. R. Burch, E. M. Clarke and D. E. Long, "Representing circuits more efficiently in symbolic model checking," in *Proc. ACM/IEEE 28th Design Automation Conf.*, June 1991, pp. 403–407.

[BURC91b] J. R. Burch, "Using BDDs to verify multipliers," in *Proc. ACM/IEEE 28th Design Automation Conf.*, June 1991, pp. 408–412.

[BUTL87] K. M. Butler, "Binary decision diagrams: Implementation, operations, and applications," Master's Report, The University of Texas at Austin, Dec. 1987.

[BUTL88] K. M. Butler, "Experimentation with new methods in ATPG," Progress Report for Individual Research Course, The University of Texas at Austin, Aug. 1988.

[BUTL89] K. M. Butler, D. E. Ross and M. R. Mercer, "An exact computation of OBDD size for classes of switching functions and combinational circuits," unpublished manuscript, May 1989.

[BUTL90a] K. M. Butler and M. R. Mercer, "On evaluating target fault models and non-target fault detection," unpublished manuscript, Apr. 1990.

[BUTL90b] K. M. Butler and M. R. Mercer, "The influences of fault type and topology on fault model performance and the implications to test and testable design," in *Proc. ACM/IEEE 27th Design Automation Conf.*, June 1990, pp. 673–678.

[BUTL90c] K. M. Butler, R. Kapur and M. R. Mercer, "The roles of controllability and observability in test," unpublished manuscript, July 1990.

[BUTL91a] K. M. Butler, D. E. Ross, R. Kapur and M. R. Mercer, "Heuristics to compute variable orderings for efficient manipulation of ordered binary decision diagrams," in *Proc. ACM/IEEE 28th Design Automation Conf.*, June 1991, pp. 417–420.

[BUTL91b] K. M. Butler and M. R. Mercer, "Quantifying non-target defect detection by target fault test sets," in *Proc. 2nd European Test Conf.*, Apr. 1991, pp. 91–100.

[CASE75] G. R. Case, "A statistical method for test sequence evaluation," in *Proc. ACM/IEEE 12th Design Automation Conf.*, June 1975, pp. 257–260.

[CERN79] E. Cerny, D. Mange and E. Sanchez, "Synthesis of minimal binary decision trees," *IEEE Trans. Comput.*, vol. C-28, no. 7, pp. 472–482, July 1979.

[CERN85] E. Cerny and J. Gecsei, "Simulation of MOS circuits by decision diagrams," *IEEE Trans. Computer-Aided Design*, vol. CAD-4, no. 4, pp. 685–693, Oct. 1985.

[CERN90] E. Cerny and C. Mauras, "Tautology checking using cross-controllability and cross-observability relations," in *Dig. Technical Papers, 1990 IEEE Int. Conf. CAD*, Nov. 1990, pp. 34–37.

[CHA74] C. Cha and G. Metze, "Multiple fault diagnosis in combinational networks," in *Proc. 12th Annual Allerton Conf. on Circuits and System Theory*, Oct. 1974, pp. 244–253.

[CHAN86] H. P. Chang, W. A. Rogers and J. A. Abraham, "Structured functional level test generation using binary decision diagrams," in *Proc. 1986 IEEE Int. Test Conf.*, Sept. 1986, pp. 97–104.

[CHEN91] K-C. Chen, Y. Matsunaga and S. Muroga, "A resynthesis approach for network optimization," in *Proc. ACM/IEEE 28th Design Automation Conf.*, June 1991, pp. 458–463.

[CHO89] K. Cho and R. E. Bryant, "Test pattern generation for sequential MOS circuits by symbolic fault simulation ," in *Proc. ACM/IEEE 26th Design Automation Conf.*, June 1989, pp. 418–423.

[CLAR90] E. M. Clarke, J. R. Burch, K. L. McMillan and D. L. Dill, "Sequential circuit verification using symbolic model checking," in *Proc. ACM/IEEE 27th Design Automation Conf.*, June 1990, pp. 46–51.

[CÔRT86] M. L. Côrtes and E. J. McCluskey, "An experiment on intermittent-failure mechanisms," in *Proc. 1986 IEEE Int. Test Conf.*, Sept. 1986, pp. 435–442.

[COUD90] O. Coudert and J. C. Madre, "A unified framework for the formal verification of sequential circuits," in *Dig. Technical Papers, 1990 IEEE Int. Conf. CAD*, Nov. 1990, pp. 126–129.

[COX88] H. Cox and J. Rajski, "A method of fault analysis for test generation and fault diagnosis," *IEEE Trans. Computer-Aided Design*, vol. CAD-7, no. 7, pp. 813–833, July 1988.

[DAVI76] R. David and G. Blanchet, "About random fault detection of combinational networks," *IEEE Trans. Comput.*, vol. C-25, no. 6, pp. 659–664, June 1976.

[DAVI81] R. David and P. Thévenod-Fosse, "Random testing of integrated circuits," *IEEE Trans. Inst. Meas.*, vol. IM-30, no. 1, pp. 20–25, Mar. 1981.

[DEGU91] Y. Deguchi, N. Ishiura and S. Yajima, "Probabilistic CTSS: Analysis of timing error probability in asynchronous logic circuits," in *Proc. ACM/IEEE 28th Design Automation Conf.*, June 1991, pp. 650–655.

[DEO74] N. Deo, *Graph theory with applications to engineering and computer science.* Englewood Cliffs, NJ: Prentice-Hall, 1974.

[DEVA89] S. Devadas, "Optimal layout via Boolean satisfiability," in *Dig. Technical Papers, 1989 IEEE Int. Conf. CAD*, Nov. 1989, pp. 294–297.

[DEVA91] S. Devadas, K. Keutzer and S. Malik, "A synthesis-based test generation and compaction algorithm for multifaults," in *Proc. ACM/IEEE 28th Design Automation Conf.*, June 1991, pp. 359–365.

[EICH77] E. B. Eichelberger and T. W. Williams, "A logic design structure for LSI testability," in *Proc. ACM/IEEE 14th Design Automation Conf.*, June 1977, pp. 462–468.

[EICH83] E. B. Eichelberger and E. Lindbloom, "Random-pattern coverage enhancement and diagnosis for LSSD logic self-test," *IBM J. Res. Develop.*, vol. 27, no. 3, pp. 265–272, May 1983.

[ELDR59] R. D. Eldred, "Test routines based on symbolic logical statements," *J. Assoc. Comput. Mach.*, vol. 6, no. 1, pp. 33–36, 1959.

[ERCO91] S. Ercolani and G. De Micheli, "Technology mapping for electrically programmable gate arrays," in *Proc. ACM/IEEE 28th Design Automation Conf.*, June 1991, pp. 239–239.

[FERG88] F. J. Ferguson and J. P. Shen, "Extraction and simulation of realistic CMOS faults using inductive fault analysis ," in *Proc. 1988 IEEE Int. Test Conf.*, Sept. 1988, pp. 475–484.

[FERR85] A. V. Ferris-Prabhu, "Modeling the critical area in yield forecasts," *IEEE J. Solid-State Circuits*, vol. SC-20, no. 4, pp. 874–880, Aug. 1985.

[FRID74] M. Fridrich and W. A. Davis, "Minimal fault tests for combinational networks," *IEEE Trans. Comput.*, vol. C-23, no. 8, pp. 850–859, Aug. 1974.

[FRIE90] S. J. Friedman and K. J. Supowit, "Finding the optimal variable ordering for binary decision diagrams," *IEEE Trans. Comput.*, vol. C-39, no. 5, pp. 710–713, May 1990, see also, *Proc. ACM/IEEE 24th Design Automation Conf.*, pp. 348-356, June 1987.

[FUJI88] M. Fujita, H. Fujisawa and N. Kawato, "Evaluation and improvements of Boolean comparison method based on binary decision diagrams," in *Dig. Technical Papers, 1988 IEEE Int. Conf. CAD*, Nov. 1988, pp. 2–5.

[FUJI90] M. Fujita, Y. Matsunaga and T. Kakuda, "Automatic and semiautomatic verification of switch-level circuits with temporal logic and binary decision diagrams," in *Dig. Technical Papers, 1990 IEEE Int. Conf. CAD*, Nov. 1990, pp. 37–40.

[FUJI83] H. Fujiwara and T. Shimono, "On the acceleration of test generation algorithms," *IEEE Trans. Comput.*, vol. C-32, no. 12, pp. 1137–1144, Dec. 1983.

[FUJI90] H. Fujiwara, "Computational complexity of controllability / observability problems for combinational circuits," *IEEE Trans. Comput.*, vol. C-39, no. 6, pp. 762–767, June 1990, see also, *Dig. Papers, 18th Int. Symp. Fault-Tolerant Comp.*, pp. 64-69, June 1988.

[GAED88] R. K. Gaede, D. E. Ross, M. R. Mercer and K. M. Butler, "CATAPULT: Concurrent automatic testing allowing parallelization and using limited topology," in *Proc. ACM/IEEE 25th Design Automation Conf.*, June 1988, pp. 597–600.

[GALI80] J. Galiay, Y. Crouzet and M. Vergniault, "Physical versus logical fault models MOS LSI circuits: Impact on their testability," *IEEE Trans. Comput.*, vol. C-29, no. 6, pp. 527–531, June 1980.

[GIRA90] J. Giraldi and M. L. Bushnell, "EST: The new frontier in automatic test-pattern generation," in *Proc. ACM/IEEE 27th Design Automation Conf.*, June 1990, pp. 667–672.

[GLOV88] C. T. Glover and M. R. Mercer, "A method of delay fault test generation," in *Proc. ACM/IEEE 25th Design Automation Conf.*, June 1988, pp. 90–95.

[GOEL81] P. Goel, "An implicit enumeration algorithm to generate tests for combinational logic circuits," *IEEE Trans. Comput.*, vol. C-30, no. 3, pp. 215–222, Mar. 1981.

[GOLD79] L. H. Goldstein, "Controllability/Observability analysis of digital circuits," *IEEE Trans. Circ. Syst.*, vol. CAS-26, no. 9, pp. 685–693, Sept. 1979.

[GRIF80] D. Griffin, "Estimation of DC stuck-fault quality levels through application of a mixed Poisson model," in *Proc. 1980 IEEE Int. Conf. Circuits and Computers*, June 1980, pp. 1099–1102.

[HAMI91] N. B. Hamida and B. Kaminska, "Hierarchical functional level testability analysis," in *Proc. 2nd European Test Conf.*, Apr. 1991, pp. 327–332.

[HARR80] R. A. Harrison, R. W. Holzwarth, P. R. Motz, R. G. Daniels, J. S. Thomas and W. H. Wiemann, "Logic fault verification of LSI: How it benefits the user ," in *Proc. 1980 WESCON*, 1980.

[HAYE85] J. P. Hayes, "Fault modeling," *IEEE Design and Test of Comput.*, vol. 2, no. 2, pp. 88–95, Apr. 1985.

[HUGH86] J. L. A. Hughes and E. J. McCluskey, "Multiple stuck-at fault coverage of single stuck-at fault test sets," in *Proc. 1986 IEEE Int. Test Conf.*, Sept. 1986, pp. 368–374.

[HUNG89] C-T. Hung, K. M. Butler and M. R. Mercer, "Improved test set generation with ordered binary decision diagrams and functional decomposition," unpublished manuscript, Apr. 1989.

[IBAR75] O. H. Ibarra and S. K. Sahni, "Polynomially complete fault detection problems," *IEEE Trans. Comput.*, vol. C-24, no. 3, pp. 242–249, Mar. 1975.

[ISHI90] N. Ishiura, S. Yajima and Y. Deguchi, "Coded time-symbolic simulation using shared binary decision diagram," in *Proc. ACM/IEEE 27th Design Automation Conf.*, June 1990, pp. 130–135.

[JACO89] M. Jacomet, "FANTESTIC: Towards a powerful fault analysis and test pattern generator for integrated circuits," in *Proc. 1989 IEEE Int. Test Conf.*, Aug. 1989, pp. 633–642.

[JAIN91] A. Jain and R. E. Bryant, "Mapping switch-level simulation onto gate-level hardware accelerators," in *Proc. ACM/IEEE 28th Design Automation Conf.*, June 1991, pp. 219–222.

[JAIN84] S. K. Jain and V. D. Agrawal, "STAFAN: An alternative to fault simulation," in *Proc. ACM/IEEE 21st Design Automation Conf.*, June 1984, pp. 475–480.

[JAIN86] S. K. Jain and D. M. Singer, "Characteristics of statistical fault analysis," in *Proc. Int. Conf. Comput. Design*, Oct. 1986, pp. 24–30.

[JU91] Y-C. Ju and R. A. Saleh, "Incremental techniques for the identification of statically sensitizable critical paths," in *Proc. ACM/IEEE 28th Design Automation Conf.*, June 1991, pp. 541–546.

[KAMA74] S. Kamal and C. V. Page, "Intermittent faults: A model and a detection procedure," *IEEE Trans. Comput.*, vol. C-23, no. 7, pp. 713–719, July 1974.

[KAPU90] R. Kapur, personal communication, Nov. 1990.

[KAPU91] R. Kapur, K. M. Butler, D. E. Ross and M. R. Mercer, "On bridging fault controllability and observability and their correlations to detectability," in *2nd European Test Conf.*, Apr. 1991, pp. 333–339.

[KIRK87] T. E. Kirkland and M. R. Mercer, "A topological search algorithm for ATPG," in *Proc. ACM/IEEE 24th Design Automation Conf.*, June 1987, pp. 502–508.

[LAMO83] P. Lamoureux and V. K. Agarwal, "Non-stuck-at fault detection in nMOS circuits by region analysis," in *Proc. 1983 IEEE Int. Test Conf.*, Oct. 1983, pp. 129–137.

[LEE59] C. Y. Lee, "Representation of switching circuits by binary-decision programs," *Bell Syst. Tech. J.*, vol. 38, pp. 985–999, July 1959.

[LESS80] J. D. Lesser and J. J. Schedletsky, "An experimental delay test generation for LSI," *IEEE Trans. Comput.*, vol. C-29, no. 3, pp. 235–248, Mar. 1980.

[LIN90a] B. Lin, H'eJ. Touati and A. R. Newton, "Don't care minimization of multi-level sequential logic networks," in *Dig. Technical Papers, 1990 IEEE Int. Conf. CAD*, Nov. 1990, pp. 414–417.

[LIN90b] B. Lin and F. Somenzi, "Minimization of symbolic relations," in *Dig. Technical Papers, 1990 IEEE Int. Conf. CAD*, Nov. 1990, pp. 88–91.

[MADR88] J-C. Madre and J-P. Billon, "Proving circuit correctness using formal comparison between expected and extracted behaviour," in *Proc. ACM/IEEE 25th Design Automation Conf.*, June 1988, pp. 205–210.

[MALA86] Y. K. Malaiya, A. P. Jayasumana and R. Rajsuman, "A detailed examination of bridging faults," in *Proc. Int. Conf. Comput. Design*, Oct. 1986, pp. 78–81.

[MALI88] S. Malik, A. R. Wang, R. K. Brayton and A. L. Sangiovanni-Vincentelli, "Logic verification using binary decision diagrams in a logic synthesis environment," in *Dig. Technical Papers, 1988 IEEE Int. Conf. CAD*, Nov. 1988, pp. 6–9.

[MALI90] S. Malik, personal communication, May 1990.

[MATS90] Y. Matsunaga, M. Fujita and T. Kakuda, "Multi-level logic minimization across latch boundaries," in *Dig. Technical Papers, 1990 IEEE Int. Conf. CAD*, Nov. 1990, pp. 406–409.

[MAXW90] P. C. Maxwell and H-J. Wunderlich, personal communication, Apr. 1990.

[MAXW91] P. C. Maxwell and R. C. Aitken, personal communication, Apr. 1991.

[MCCA63] J. McCarthy, "A basis for a mathematical theory of computation," *Computer Programming and Formal Systems* (Studies in Logic and the Foundations of Mathematics). Amsterdam: North-Holland, pp. 33–70, 1963.

[MCCL56] E. J. McCluskey, "Detection of group invariance or total symmetry of a Boolean function," *Bell Syst. Tech. J.*, vol. 35, pp. 1445–1453, Nov. 1956.

[MCCL71] E. J. McCluskey and F. W. Clegg, "Fault equivalence in combinational logic networks," *IEEE Trans. Comput.*, vol. C-20, no. 11, pp. 1286–1293, Nov. 1971.

[MCCL86] E. J. McCluskey, *Logic Design Principles with Emphasis on Testable Semicustom Circuits.* Englewood Cliffs, NJ: Prentice-Hall, 1986.

[MCCL88] E. J. McCluskey and F. Buelow, "IC quality and test transparency," in *Proc. 1988 IEEE Int. Test Conf.*, Sept. 1988, pp. 295–301.

[MCGE89] P. C. McGeer and R. K. Brayton, "Efficient algorithms for computing the longest viable path in a combinational network," in *Proc. ACM/IEEE 26th Design Automation Conf.*, June 1989, pp. 561–567.

[MEI74] K. C. Y. Mei, "Bridging and stuck-at faults," *IEEE Trans. Comput.*, vol. C-23, no. 7, pp. 720–727, July 1974.

[MENO65] P. R. Menon, "A simulation program for logic networks," Bell Telephone Labs. Internal Technical Memorandum, No. MM65-1271-3, Mar. 1965.

[MERC84] M. R. Mercer and B. Underwood, "Correlating testability with fault detection," in *Proc. 1984 IEEE Int. Test Conf.*, Oct. 1984, pp. 697–704.

[MIDK89] S. F. Midkiff and W-Y. Koe, "Test effectiveness metrics for CMOS faults," in *Proc. 1989 IEEE Int. Test Conf.*, Aug. 1989, pp. 653–659.

[MILL88] S. D. Millman and E. J. McCluskey, "Detecting bridging faults with stuck-at test sets," in *Proc. 1988 IEEE Int. Test Conf.*, Sept. 1988, pp. 773–783.

[MILL89] S. D. Millman and E. J. McCluskey, "Detecting stuck-open faults with stuck-at test sets," in *Proc. 1989 IEEE Custom Integrated Circ. Conf.*, May 1989, pp. 22.3.1–22.3.4.

[MINA90] S-ichi Minato, N. Ishiura and S. Yajima, "Shared binary decision diagram with attributed edges for efficient boolean functional manipulation," in *Proc. ACM/IEEE 27th Design Automation Conf.*, June 1990, pp. 52–57.

[MOOR79] G. E. Moore, "Are we really ready for VLSI^2?," in *Proc. Caltech Conference on VLSI*, Jan. 1979, pp. 3–14.

[MORE82] B. M. E. Moret, "Decision trees and diagrams," *ACM Computing Surveys*, vol. 14, no. 4, pp. 593–623, Dec. 1982.

[NAIR86] R. Nair and D. Brand, "Construction of optimal DCVS trees," IBM Res. Rep. RC 11863, Mar. 1986.

[NAJM91] F. N. Najm, "Transition density, a stochastic measure of activity in digital circuits," in *Proc. ACM/IEEE 28th Design Automation Conf.*, June 1991, pp. 644–649.

[NICK80] V. V. Nickel, "VLSI - The inadequacy of the stuck at fault model," in *Proc. 1980 IEEE Test Conf.*, Nov. 1980, pp. 378–381.

[NIGH89] P. Nigh and W. Maly, "Layout-driven test generation," in *Dig. Technical Papers, 1989 IEEE Int. Conf. CAD*, Nov. 1989, pp. 154–157.

[OCHI91] H. Ochi, N. Ishiura and S. Yajima, "Breadth-first manipulation of SBDD of Boolean functions for vector processing," in *Proc. ACM/IEEE 28th Design Automation Conf.*, June 1991, pp. 413–416.

[OHMU90] M. Ohmura, H. Yasuura and K. Tamura, "Extraction of functional information from combinational circuits," in *Dig. Technical Papers, 1990 IEEE Int. Conf. CAD*, Nov. 1990, pp. 176–179.

[OSBU88] C. M. Osburn, H. Berger, R. P. Donovan and G. W. Jones, "The effects of contamination on semiconductor manufacturing yield," *J. Environmental Sciences*, vol. 31, no. 2, pp. 45–57, Mar.-Apr. 1988.

[PANC90] A. Pancholy, J. Rajski and L. J. McNaughton, "Empirical failure analysis and validation of fault models in CMOS," in *Proc. 1990 IEEE Int. Test Conf.*, Sept. 1990, pp. 938–947.

[PARK87] E. S. Park and M. R. Mercer, "Robust and nonrobust tests for path delay faults in a combinational circuit," in *Proc. 1987 IEEE Int. Test Conf.*, Sept. 1987, pp. 1027–1034.

[PARK89] E. S. Park, M. R. Mercer and T. W. Williams, "A statistical model for delay-fault coverage," *IEEE Design and Test of Comput.*, vol. 6, no. 1, pp. 45–55, Feb. 1989.

[PARK75] K. P. Parker and E. J. McCluskey, "Analysis of logic circuits with faults using input signal probabilities," *IEEE Trans. Comput.*, vol. C-24, no. 5, pp. 573–578, May 1975.

[PAYN77] H. J. Payne and W. S. Meisel, "An algorithm for constructing optimal binary decision trees," *IEEE Trans. Comput.*, vol. C-26, no. 9, pp. 905–916, Sept. 1977.

[RATI83] I. M. Ratiu, "VICTOR: Global redundancy identification and test generation for VLSI circuits," PhD Dissertation, University of California - Berkeley, May 1983.

[ROSS89] D. E. Ross, personal communication, Aug. 1989.

[ROSS90] D. E. Ross, "Functional calculations using ordered partial multi decision diagrams," PhD Dissertation, The University of Texas at Austin, Aug. 1990.

[ROSS91a] D. E. Ross, K. M. Butler and M. R. Mercer, "Exact ordered binary decision diagram size when representing classes of symmetric functions," *Journal of Electronic Testing: Theory and Practice (JETTA)*, vol. 2, no. 3, 1991, to appear.

[ROSS91b] D. E. Ross, K. M. Butler, R. Kapur and M. R. Mercer, "Fast functional evaluation of candidate OBDD variable orderings," in *Proc. 3rd European Design Automation Conf.*, Feb. 1991, pp. 4–10.

[ROTH66] J. P. Roth, "Diagnosis of automata failures: A calculus and a method," *IBM J. Res. Develop.*, vol. 10, pp. 278–291, July 1966.

[SATO90] H. Sato, Y. Yasue, Y. Matsunaga and M. Fujita, "Boolean resubstitution with permissible functions and binary decision diagrams," in *Proc. ACM/IEEE 27th Design Automation Conf.*, June 1990, pp. 284–289.

[SAVI80] J. Savir, "Syndrome-testable design of combinational circuits," *IEEE Trans. Comput.*, vol. C-29, no. 6, pp. 442–451, June 1980.

[SAVI83] J. Savir, "Good controllability and observability do not guarantee good testability," *IEEE Trans. Comput.*, vol. C-32, no. 12, pp. 1198–1200, Dec. 1983.

[SAVI84] J. Savir, G. H. Ditlow and P. H. Bardell, "Random pattern testability," *IEEE Trans. Comput.*, vol. C-33, no. 1, pp. 79–90, Jan. 1984.

[SAVI90] J. Savir, "AC product defect level and yield loss," in *Proc. 1990 IEEE Int. Test Conf.*, Sept. 1990, pp. 726–738.

[SCHE72] D. R. Schertz and G. Metze, "A new representation for faults in combinational digital circuits," *IEEE Trans. Comput.*, vol. C-21, no. 8, pp. 858–866, Aug. 1972.

[SCHU88a] M. H. Schulz and E. Auth, "Advanced automatic test pattern generation and redundancy identification techniques," in *Dig. Papers, 18th Int. Symp. Fault-Tolerant Comp.*, June 1988, pp. 30–35.

[SCHU88b] M. H. Schulz, E. Trischler and T. M. Sarfert, "SOCRATES: A highly efficient automatic test pattern generation system," *IEEE Trans. Computer-Aided Design*, vol. CAD-7, no. 1, pp. 126–137, Jan. 1988.

[SETH73] S. C. Seth, "Distance measures on detection test sets and their applications," in *Dig. Papers, 3rd Int. Symp. Fault-Tolerant Comput.*, June 1973, pp. 101–104.

[SETH84] S. C. Seth and V. D. Agrawal, "Characterizing the LSI yield equation from wafer test data," *IEEE Trans. Computer-Aided Design*, vol. CAD-3, no. 2, pp. 123–126, Apr. 1984.

[SETH90] S. C. Seth, V. D. Agrawal and H. Farhat, "A statistical theory of digital circuit testability," *IEEE Trans. Comput.*, vol. C-39, no. 4, pp. 582–586, Apr. 1990, see also, *Proc. 1st Euro. Test Conf.*, pp. 139-143, Apr. 1989, and *Proc. Int. Conf. Comput. Design*, pp. 58-61, Oct. 1988.

[SHAN38] C. E. Shannon, "A symbolic analysis of relay and switching circuits," *Trans. AIEE*, vol. 57, pp. 713–723, 1938.

[SHEN85] J. P. Shen, W. Maly and F. J. Ferguson, "Inductive fault analysis of MOS integrated circuits," *IEEE Design and Test of Comput.*, vol. 2, no. 6, pp. 13–26, Dec. 1985.

[SILB87] G. M. Silberman and I. Spillinger, "Test generation using functional fault simulation and the difference fault model," in *Proc. 1987 IEEE Int. Test Conf.*, Sept. 1987, pp. 400–407.

[SILB90] G. M. Silberman and I. Spillinger, "Using functional fault simulation and the difference fault model to estimate implementation fault coverage," *IEEE Trans. Computer-Aided Design*, vol. CAD-9, no. 12, pp. 1335–1343, Dec. 1990.

[SILV85] M. Silva and R. David, "Binary-decision graphs for implementation of Boolean functions," *IEE Proceedings Pt. E*, vol. 3, pp. 175–185, May 1985.

[SIMO88] H. Simonis, N. Nguyen and M. Dincbas, "Verification of digital circuits using CHIP," *The Fusion of Hardware Design and Verification*, G. J. Milne, Ed. Amsterdam: Elsevier Science Publishers B.V. (North-Holland), pp. 421–442, 1988.

[SMIT85] G. L. Smith, "Model for delay faults based upon path," in *Proc. 1985 IEEE Int. Test Conf.*, Nov. 1985, pp. 342–349.

[SODE89] J. M. Soden, R. K. Treece, M. R. Taylor and C. F. Hawkins, "CMOS IC stuck-open fault electrical effects and design considerations," in *Proc. 1989 IEEE Int. Test Conf.*, Aug. 1989, pp. 423–430.

[SRC85] SRC, "Guidelines for research proposals, criteria for funding," Semiconductor Research Corporation, Aug. 1985.

[SRIN90] A. Srinivasan, T. Kam, S. Malik and R. K. Brayton, "Algorithms for discrete function manipulation," in *Dig. Technical Papers, 1990 IEEE Int. Conf. CAD*, Nov. 1990, pp. 92–95.

[STAN88] D. Stannard and B. Kaminska, "Detection of hard faults in a combinational circuit using budget constraints," in *Proc. 1988 IEEE Int. Test Conf.*, Sept. 1988, p. 999.

[STAP75] C. H. Stapper, "On a composite model to the I. C. yield problem," *IEEE J. Solid-State Circuits*, vol. SC-10, no. 6, pp. 537–539, Dec. 1975.

[STAP83] C. H. Stapper, "Modeling of integrated circuit defect sensitivities," *IBM J. Res. Develop.*, vol. 27, no. 6, pp. 549–557, Nov. 1983.

[STAP84] C. H. Stapper, "Modeling of defects in integrated circuit photolithographic patterns," *IBM J. Res. Develop.*, vol. 28, no. 4, pp. 549–557, July 1984.

[STOR90] T. Storey, W. Maly, J. Andrews and M. Miske, "Assessing CMOS test quality: Theory vs. practice," in *TECHCON '90 Extended Abstract Volume*, Oct. 1990, pp. 387–390.

[STOR91] T. Storey, W. Maly, J. Andrews and M. Miske, "Comparing stuck fault and current testing via CMOS chip test," in *Proc. 2nd European Test Conf.*, Apr. 1991, pp. 149–156.

[SUPO86] K. J. Supowit and S. J. Friedman, "A new method for verifying sequential circuits," in *Proc. ACM/IEEE 23rd Design Automation Conf.*, June 1986, pp. 200–207.

[THAT80] S. M. Thatte and J. A. Abraham, "Test generation for microprocessors," *IEEE Trans. Comput.*, vol. C-29, no. 6, pp. 429–441, June 1980.

[TIMO83] C. Timoc, M. Buehler, T. Griswold, C. Pina, F. Stott and L. Hess, "Logical models of physical failures," in *Proc. 1983 IEEE Int. Test Conf.*, Oct. 1983, pp. 546–553.

[TOUA90] H. J. Touati, H. Savoj, B. Lin, R. K. Brayton and A. Sangiovanni-Vincentelli, "Implicit state enumeration of finite state machines using BDD's," in *Dig. Technical Papers, 1990 IEEE Int. Conf. CAD*, Nov. 1990, pp. 130–133.

[WADS78a] R. L. Wadsack, "Fault modeling and logic simulation of CMOS and MOS integrated circuits," *Bell Syst. Tech. J.*, vol. 57, no. 5, pp. 1449–1474, May-June 1978.

[WADS78b] R. L. Wadsack, "Fault coverage in digital integrated circuits," *Bell Syst. Tech. J.*, vol. 57, no. 5, pp. 1475–1488, May-June 1978.

[WADS81] R. L. Wadsack, "VLSI: How much fault coverage is enough?," in *Proc. 1981 IEEE Int. Test Conf.*, Sept. 1981, pp. 547–554.

[WAIC85] J. A. Waicukauski, E. B. Eichelberger, D. Forlenza, E. Lindbloom and T. McCarthy, "Fault simulation for structured VLSI design," *VLSI Systems Design*, vol. 6, no. 12, pp. 20–32, Dec. 1985.

[WILL90] R. H. Williams and C. F. Hawkins, "Errors in testing," in *Proc. 1990 IEEE Int. Test Conf.*, Sept. 1990, pp. 1018–1027.

[WILL81] T. W. Williams and N. C. Brown, "Defect level as a function of fault coverage," *IEEE Trans. Comput.*, vol. C-30, no. 12, pp. 987–988, Dec. 1981.

[WOOD87] B. W. Woodhall, B. D. Newman and A. G. Sammuli, "Empirical results on undetected CMOS stuck-open failures," in *Proc. 1987 IEEE Int. Test Conf.*, Sept. 1987, pp. 166–170.

Index